Signal Transduct

2

The Practical Approach Series

SERIES EDITORS

D. RICKWOOD
Department of Biology, University of Essex
Wivenhoe Park, Colchester, Essex CO4 3SQ, UK

B. D. HAMES
Department of Biochemistry and Molecular Biology, University of Leeds
Leeds LS2 9JT, UK

Affinity Chromatography
Anaerobic Microbiology
Animal Cell Culture
   (2nd Edition)
Animal Virus Pathogenesis
Antibodies I and II
Biochemical Toxicology
Biological Membranes
Biomechanics—Materials
Biomechanics—Structures and
   Systems
Biosensors
Carbohydrate Analysis
Cell Growth and Division
Cellular Calcium
Cellular Neurobiology
Centrifugation (2nd Edition)
Clinical Immunology
Computers in Microbiology
Crystallization of Proteins and
   Nucleic Acids
Cytokines
The Cytoskeleton

Diagnostic Molecular Pathology
   I and II
Directed Mutagenesis
DNA Cloning I, II, and III
Drosophila
Electron Microscopy in
   Biology
Electron Microscopy in
   Molecular Biology
Enzyme Assays
Essential Molecular Biology
   I and II
Fermentation
Flow Cytometry
Gel Electrophoresis of Nucleic
   Acids (2nd Edition)
Gel Electrophoresis of Proteins
   (2nd Edition)
Genome Analysis
HPLC of Macromolecules
HPLC of Small Molecules
Human Cytogenetics I and II
   (2nd Edition)

Signal Transduction
A Practical Approach

Edited by
G. MILLIGAN

Molecular Pharmacology Group,
Departments of Biochemistry and Pharmacology,
University of Glasgow,
Glasgow G12 8QQ
Scotland, UK

OXFORD UNIVERSITY PRESS
Oxford New York Tokyo

Oxford University Press, Walton Street, Oxford OX2 6DP

Oxford New York Toronto
Delhi Bombay Calcutta Madras Karachi
Petaling Jaya Singapore Hong Kong Tokyo
Nairobi Dar es Salaam Cape Town
Melbourne Auckland
and associated companies in
Berlin Ibadan

Oxford is a trade mark of Oxford University Press

A Practical Approach ⬡ is a registered trade mark
of the Chancellor, Masters, and Scholars of the University of Oxford
trading as Oxford University Press

Published in the United States
by Oxford University Press, New York

A catalogue record for this book is available from the British Library

Library of Congress Cataloging in Publication Data
Signal transduction: a practical approach/edited by G. Milligan.
p. cm.—(The Practical approach series)
Includes bibliographical references.
1. G proteins—Research—Methodology. 2. Cellular signal
transduction—Research—Methodology. I. Milligan, Graeme.
II. Series.
QP552.G16S54 1992 574.87'6—dc20 92–195

ISBN 0–19–963296–0 (hb.)
ISBN 0–19–963295–2 (pb.)

Typeset by Cambrian Typesetters, Frimley, Surrey
Printed by Information Press Ltd., Eynsham, Oxon.

This book is dedicated to Phil Godfrey (1957–1991) who died during its production.

Preface

Cellular trans-plasma membrane signalling processes are amongst the most highly studied processes in the biological sciences. A vectorial transduction cascade is required to convert extracellular information into an intracellular message which will in turn lead to a physiological response. In the case of G-protein mediated signalling the traditional view is that only receptors of the 'seven transmembrane element' family cause the activation of G-proteins upon agonist occupancy of the receptor. Such orthodoxy has been challenged but it is clear that this class of receptors are, at least numerically, by far the most important regulators of G-proteins. As such, the first chapter describes what information can be gained about interactions of receptors and G-proteins by analysing ligand binding to such receptors. Chapters 2 and 3 provide basic assay and detection systems for G-proteins and Chapter 3 also demonstrates how mutagenesis can be used to define functional domains of G-proteins. Chapters 4–7 describe assay methods for effector enzymes which are regulated subsequent to receptor activation of a G-protein and for the detection of the intracellular signals so generated while Chapter 8 indicates the types of strategies being used by electrophysiologists to study G-protein regulation of ion channels.

Although a book of this nature can provide the details only of commonly used practical approaches it is hoped that it may provide a useful guide to the newcomer to the field as well as to allow workers already in the field to expand the range of approaches in use within their laboratory as well as to point out some of the practical difficulties and limitations involved.

Glasgow   G. M.
December 1991

Contents

3. Reconstitution of cyc⁻ membranes by in vitro translated $G_s\alpha$: a model for studying functional domains of $G_s\alpha$ subunit

Yves Audigier

6. Phosphatidylcholine hydrolysis by phospholipases C and D

R. W. Bonser and N. T. Thompson

7. The determination of phospholipase A₂ activity in stimulated cells

Michael J. O. Wakelam and Susan Currie

8. Electrophysiological approaches to G-protein function

I. McFadzean and D. A. Brown

Contributors

LINDA M. ALLAN
Department of Biochemistry, University of Cambridge, Tennis Court Road, Cambridge CB2 1QW.

YVES AUDIGIER
Centre CNRS-INSERM de Pharmacologie-Endocrinologie, F-34094 Montpelier, France.

R. W. BONSER
Department of Biochemical Sciences, Wellcome Research Laboratories, Beckenham, Kent BR3 3BS.

D. A. BROWN
Department of Pharmacology, University College London, Gower St, London WC1E 6BT.

SUSAN CURRIE
Molecular Pharmacology Group, Department of Biochemistry, University of Glasgow, Glasgow G12 8QQ.

RICHARD FARNDALE
Department of Biochemistry, University of Cambridge, Tennis Court Road, Cambridge CB2 1QW.

PHILIP P. GODFREY
Glaxo Institute for Molecular Biology, Geneva, Switzerland.

D. GRAESER
Department of Pharmacology, University of Michigan, Ann Arbour, Michigan 48109–0626, USA.

I. McFADZEAN
Pharmacology Group, Biomedical Sciences Division, King's College London, Manresa Road, London SW3 6LX.

FERGUS R. McKENZIE
Centre de Biochimie, CNRS, Parc Valrose, Nice 06034, France.

B. RICHARD MARTIN
Department of Biochemistry, University of Cambridge, Tennis Court Road, Cambridge CB2 1QW.

R. R. NEUBIG
Department of Pharmacology, University of Michigan, Ann Arbour, Michigan 48109–0626, USA.

Contributors

N. T. THOMPSON
Department of Biochemical Sciences, Wellcome Research Laboratories, Beckenham, Kent BR3 3BS.

MICHAEL J. O. WAKELAM
Molecular Pharmacology Group, Department of Biochemistry, University of Glasgow, Glasgow G12 8QQ.

Abbreviations

App[NH]p	adenosine 5'(β,γ,imido) diphosphate
BSA	bovine serum albumin
cAMP	3'5' cyclic adenosine monophosphate
cDNA	complementary DNA
cGMP	3'5' cyclic guanosine monophosphate
CDP	cytidine diphosphate
CDP-DG	cytidine diphosphate-diacylglycerol
Chaps	cholamidopropyl-dimethylammonio-propane sulfonate
CTP	cytidine triphosphate
DAG	diacylglycerol
DGK	diacylglycerol kinase
DRG	diradylglycerols
DTPA	diethylenetriaminepentaacetic acid
DTT	dithiothreitol
DMSO	dimethylsulfoxide
EDTA	ethylenedinitrilo-tetraacetic acid (disodium salt)
EGTA	ethylene glycol-bis (β-aminoethyl ether) N,N,N',N'-tetraacetic acid
FFA	free fatty acid
FREON	1,1,2-trichlorotrifluoroethane
G-protein	guanine nucleotide binding protein
GDP	guanosine diphosphate
GDPβS	guanosine-5'-O(2-thiodiphosphate)
GTP	guanosine triphosphate
GTPγS	guanosine 5'-O(3-thiotriphosphate)
G_i	guanine nucleotide binding protein which regulates inhibition of adenylate cyclase
G_o	'other' G-protein
Gpp[NH]p	guanylylimidodiphosphate
G_s	guanine nucleotide binding protein which regulates the stimulation of adenylate cyclase
HBH	Hepes-buffered Hanks' balanced salt solution
Hepes	N-2-hyrodroxyethylpiperazine-N'-2-ethane-sulfonic acid
HPLC	high-performance liquid chromatography
IBMX	isobutylmethylxanthine
IC_{50}	concentration inhibiting 50% of response
IHYP	iodohydroxylpindolol
IP	inositol phosphate (this definition implies no attempt to identify the particular species)
IP_2	inositol bisphosphate
IP_3	inositol trisphosphate (the numbering configuration of the position of the phosphates on the inositol ring defines the identity of the isomeric form).
IP_4	inositol tetrakisphosphate

Mops	morpholinopropane sulfonic acid
NDGA	nordihydroguaiaretic acid
NTP	nucleoside triphosphate
PA	phosphatidic acid
PBS	phosphate-buffered saline
PBut	phosphatidylbutanol
PC	phosphatidylcholine
PCA	perchloric acid
PI	phosphatidylinositol
PIC	phosphoinositide-specific phospholipase C
PIP	phosphatidylinositol 5-monophosphate
PIP_2	phosphatidylinositol 4,5-bisphosphate
PKC	protein kinase C
PLA_1	phospholipase A_1
PLA_2	phospholipase A_2
PLC	phospholipase C
PLD	phospholipase D
PMA	phorbol myristate acetate (also called TPA)
p[NH]ppG	see Gpp[NH]p
p[NH]ppA	see App[NH]p
PS	phosphatidylserine
R	receptor
RG	receptor/G-protein
RIA	radioimmunoassay
SAG	1-stearoyl-2-arachidonyl-*sn*-glycerol
TLC	thin-layer chromatography
Tris	2-amino-2-(hydroxymethyl)-1,3-propandiol
UK14304	5-bromo-6-*N*-(2-4,5,-dihydroimidazolyl) quinoxaline
UTP	uridine monophosphate

Methods for the study of receptor/ G-protein interactions

D. GRAESER and R. R. NEUBIG

1. Introduction

Receptor conformation and guanine nucleotide regulatory protein (G-protein) coupling have received widespread attention due to their role in the mechanism of action of many hormones, neurotransmitters, and autocoids. There are tremendous numbers of different receptors and G-proteins that utilize similar mechanisms (1–3). Thus, the experimental approaches to the study of one system will often yield methods that can be applied in a relatively simple manner to other receptor/G-protein (RG) systems. Because of the tremendous diversity of receptor and G-protein systems, the protocols presented here cannot encompass all possibilities but they may serve as a guide to developing methods for the specific system being studied. Since receptors and G-proteins have only been demonstrated to interact in membranes, either native or reconstituted, this chapter will focus on methods applicable in a membrane environment.

There are many experimental complexities in the study of membrane proteins, so careful planning is critical to the successful outcome of an experiment. Several factors are particularly important, including the source of biological material, the type of assay and the method of data analysis. This chapter will present examples that illustrate practical approaches to solving these problems.

Currently, a wide variety of experimental procedures are available to explore RG interactions. These include radioligand binding, enzymatic assays (such as GTP hydrolysis and second messenger production), modification of RG coupling by exogenous agents and more recently spectroscopic methods. Preparation of the material used in these experiments is also critical, since the method of preparation determines what type(s) of experiments can be performed. In this chapter, particular attention will be given to the following procedures:

- preparing plasma membranes from tissue culture and primary tissue

- agonist binding for detection of RG coupling
- modelling the completed data

2. Methods for preparing material

The biological source of material and the method of preparation are critical for planning experiments. They must be considered in terms of cost, experimental limitations, difficulty of preparation, and ease of use. Some methods of preparation include plasma membranes from cells, reconstitution, and subcellular fractionation. Each has advantages and disadvantages and the method of preparation must be tailored to the experimental protocol. Some considerations include: stability of membrane protein activity (i.e. GTPase activity, cyclic AMP production, etc.), necessity of eliminating endogenous contaminants such as guanine nucleotides, and diffusion barriers, such as the plasma membrane.

2.1 Preparing plasma membranes

Purified plasma membranes can be prepared from primary tissue, such as platelets, or from cells grown in culture, such as NG108–15 cells. These membranes are useful for both ligand-binding studies and functional assays. The type of preparation varies widely. The details of a protocol are often applicable only to one source of material, although the basic steps are more generally useful. The overall goal of membrane preparation is to disrupt the cell so that the plasma membrane forms fragments which contain the receptor and G-proteins of interest. The plasma membranes may be purified further or the crude homogenate used. Different assays may require one or the other type of preparation (see *Protocol 1*). The protocols given here are intended as examples; the reader is advised to examine the literature for protocols specific for the type of tissue used. If none are available, *Protocols 1* and *2* may be used as general guides to the development of methods for the problem at hand.

Protocol 1. Gradient purification of plasma membranes from human platelets

You will need:

- human platelet concentrates (obtained from the Red Cross) in anti-coagulant
- buffer 1 (150 mM NaCl, 1 mM EDTA (ethylenedinitrilo-tetraacetic acid disodium salt), 20 mM Tris–HCl, pH 7.5)
- buffer 2 (50 mM Tris–HCl, pH 7.5, 5 mM EGTA (ethylene glycol-bis(β-aminoethyl ether) N,N,N',N'-tetraacetic acid), 0.2 M sucrose)

- buffer 3 (50 mM Tris–HCl, pH 7.6, 10 mM $MgCl_2$, 1 mM EGTA)
- 0.1 M phenylmethylsulfonyl fluoride (a serine protease inhibitor) in ethanol (1.74 mg/100 μl ethanol)
- 14.5% (w/w) sucrose solution (14.5 g sucrose, 85 g H_2O, 0.5 ml 4% NaN_3)
- 34% (w/w) sucrose solution (34 g sucrose, 65.5 g H_2O, 0.5 ml 4% NaN_3)

1. At room temperature, dilute human platelet concentrates with an equal volume of buffer 1. Centrifuge for 10 min at 200 g to remove erythrocytes.

2. Carefully remove supernatant and discard pellet. Centrifuge supernatant at 1500 g for 20 min. Gently resuspend the platelet pellet in buffer 1 (same volume as above) and pellet it again.

3. Resuspend pellet in buffer 2 (sufficient volume to form thick suspension) and quick-freeze in dry-ice/ethanol. Store at −70 °C until use.

4. Prepare a discontinuous sucrose gradient by placing 20 ml of 34% sucrose solution at the bottom of an ultracentrifuge tube, capable of withstanding 105 000 g, which is suitable for the Beckman Type 35 rotor. Layer 20 ml of 14.5% sucrose solution slowly and carefully on to the heavier sucrose solution in the tube. Handle the tubes very carefully so as not to disturb the gradients and keep them at 4 °C until use.

5. Add 2 μl 0.1 M phenylmethylsulfonyl fluoride/unit platelets to each tube of frozen platelets (final concentration approx. 10^{-4} M) and thaw at 15 °C (a cooling water bath works well). Immediately place on ice. Divide into 50-ml portions.

6. Probe sonicate each portion twice for 10 sec, using a Branson Sonifier (model W-350) cell disruptor at setting 6, chilling between sonications on ice for 30–60 sec to prevent gel formation.

7. Carefully layer 15 ml of sonicated material/tube on to these gradients, without disturbing the gradient. This step is important for removing endogenous guanine nucleotides, which can interfere with later experimental results.

8. Place tubes in Beckman Type 35 rotor and centrifuge for 3 h at 105 000 g without any brake. The rotor should be slowly accelerated and slowly decelerated, to avoid disturbing the gradient. When the spin is complete, lift the rotor carefully out of the centrifuge without jerking.

9. Remove tubes. The interface between the 14.5% and 34% sucrose contains the purified platelet membranes. Use a Pasteur pipette to carefully remove the interface, without disturbing the top layer, to prevent contaminating membranes with the soluble fraction. The soluble fraction contains a high concentration of guanine nucleotides. Gently expelling air from the pipette while penetrating the top layers of the gradient will help prevent contamination.

Protocol 1. *Continued*

10. Measure the volume of the membranes, and then dilute with two volumes of ice-cold distilled-deionized water.

11. Centrifuge for 60 min at 105 000 g. Resuspend pellet in 1 ml of buffer 3 per unit of platelets.

12. Quick-freeze aliquots in dry-ice/ethanol or liquid nitrogen and store at −70 °C until use.

This platelet membrane preparation was optimized to yield the greatest amount of high-affinity α_2-adrenergic agonist-binding (4). It has also been used very successfully for measurements of adenylyl cyclase inhibition (5) and GTPase activation (6) by α_2-adrenergic receptors. Agonist-binding to G-protein-coupled receptors exhibits high-affinity only in the absence of guanine nucleotides. Although it is not widely appreciated, both GTP *and* GDP potently reduce agonist-binding. For [^{3}H]*p*-aminoclonidine, the concentrations of nucleotide inhibiting 50% of agonist-binding are 0.1 and 0.5 µM, for GTP and GDP, respectively (Gantzos and Neubig, unpublished). *Table 1* illustrates why the gradient method is necessary to obtain optimal RG coupling as assessed by high-affinity agonist-binding. Thus, a crude platelet membrane preparation contains sufficient guanine nucleotide concentrations to greatly inhibit high-affinity agonist-binding, even after extensive washing by centrifugation in buffer (4). Even the purified plasma membrane fraction still contains sufficient guanine nucleotides to partially support α_2-adrenergic inhibition of adenylyl cyclase (5). Similar conclusions with regard to the use of gradient purification methods to eliminate guanine nucleotide contamination were reached by Ross *et al.* (7) for β-adrenergic receptors in the cultured S49 lymphoma system. This is especially important when large-scale purifications of membranes are done in which there is a lot of endogenous guanine nucleotide to remove.

A simple membrane preparation from small amounts of cultured cells will often suffice for studies of RG interactions by agonist-binding and enzymatic methods. The quicker and gentler method is not as good for removing

---

Table 1. Guanine nucleotide concentrations in platelet membrane fractions

Membrane fraction	GTP (µM)[a]	GDP (µM)[a]
Crude membranes	1.9	5.4
Washed membranes	0.9	2.5
Gradient pellet	2.1	5.7
Gradient purified plasma membranes	0.07	0.5

[a] Concentration in a binding assay with 100 pM α_2-adrenergic receptor (4).

4

nucleotides but may result in more robust activity and agonist regulation of sensitive enzymes such as adenylyl cyclase, phospholipases, and even GTPase.

Protocol 2. Preparing membranes from cells grown in culture (NG108-15)

You will need:

- NG108-15 cells grown in culture (8)
- serum-free tissue culture medium (Dulbecco's modified Eagle's minimal essential medium, supplemented with 0.1 mM hypoxanthine, 1 μM aminopterin, 16 μM thymidine)
- buffer 1 (137 mM NaCl, 5.6 mM dextrose, 1 mM EGTA, 5 mM Hepes, pH 7.4)
- buffer 2 (5 mM Tris–HCl, 5 mM $MgCl_2$, 1 mM EGTA, pH 7.5)
- buffer 3 (50 mM Tris–HCl, 10 mM $MgCl_2$, 1 mM EGTA, pH 7.6)

1. Incubate confluent NG108-15 cells (2–4 150-cm^2 flasks) in serum-free tissue culture media for 16–20 h before use, to remove endogeneous serum factors such as catecholamines.

2. Harvest cells by banging the flasks in which they were grown against a laboratory bench. The cells will become suspended in their medium, which can then be poured into centrifuge tubes. Centrifuge at 800 *g* for 5 min at 4 °C.

3. Wash once by resuspending pellet in buffer 1 and centrifuging as above.

4. Resuspend pellet in 20 ml buffer 2, and homogenize ten times. Use a motor-driven 30-ml glass-Teflon homogenizer, at 4 °C set to 600 r.p.m. Centrifuge homogenate at 1000 *g* for 5 min at 4 °C. Collect supernatant and save.

5. Repeat step 4 with the pellet. Note that during and after homogenization, the material should always be kept at 4 °C.

6. Combine supernatants, dilute to 40 ml with buffer 3, and centrifuge for 30 min at 100 000 *g*. This dilution is necessary to remove endogeneous guanine nucleotides.

7. Resuspend pellet in 5 ml buffer 3, quick-freeze in liquid nitrogen (or solid CO_2/ethanol) and store at −70 °C.

8. Thaw the membranes, and wash the membranes by diluting to 45 ml with buffer 3, and centrifuging for 30 min at 100 000 *g* before using.

[a] These cells lose their α_2 adrenergic receptors after a number of passages. Generally, cells of passage number 14–45 have an adequate number of receptors, but even within this range, cells may have variable receptor number. Older cells should be used with caution.

Some trouble-shooting tips: If a particular method for membrane preparation is unsuccessful (that is, protein yield and/or purity was low), problems often fall into one of two general extremes. The method may have been too rough, disrupting the interaction between the receptor and G-protein, or denaturing the proteins themselves. The method could also be too gentle, either not fully disrupting the cells, or inadequately separating the plasma membranes from the rest of the cellular material. The definition of 'rough' and 'gentle' depends on the cell source, and must be determined for each individual tissue. If neither of the membrane preparation methods given above works, try the following alterations:

- Use a glass homogenizer or Dounce, instead of a Teflon homogenizer. This is often less disruptive to the cells.

- Try higher ionic strength buffers (for example, buffer 3 above), or buffers with a different base, instead of the hypotonic buffers given above. These are also less disruptive to the cells.

- Include a ligand specific for the receptor in the membrane preparation, which may stabilize the receptor. Washing the membranes should remove it at the end.

- Use the crude homogenate instead of further purifying membranes. If the yield of purified membranes is extremely low, the plasma membranes may be pelleting at low speed with the crude homogenate. Also, some membranes are much more stable in a crude homogenate preparation than in a purified preparation.

2.2 Reconstitution of receptors and G-proteins

Reconstitution generally refers to purifying cellular proteins separately then recombining them in lipid vesicles. It offers the advantage of reducing a complicated biological system to a few parts. Many functional assays give best results when used with reconstituted material. Unfortunately, successful methods for reconstitution differ widely from system to system, and designing a new procedure is largely empirical. A number of factors must be optimized for successful reconstitution:

- protein purification
- lipid requirements
- detergent removal

The method of protein purification is critical for protein stability and activity, and has a profound effect on the final outcome of the reconstitution. Choosing a particular method includes selecting a detergent to solubilize the purified proteins. Once a suitable method is found, proteins must be recombined in a system which retains the activity of the protein. Some systems have specific lipid requirements, but others do not. Often, 'dirty' lipid mixtures (such as crude soybean azolectin) prove to be very successful, but in some cases a single pure lipid may be needed. Lipids can also be purified from the same source as the protein(s) and these native lipids may provide a more stable environment for some proteins (9). In order to insert the proteins into lipid vesicles, the detergent must be removed. Some detergents, such as digitonin, Lubrol or Triton, are extremely difficult to remove. The requirement to remove the detergent will affect the original choice of detergent for protein solubilization. Usually, a detergent with a high critical micellar concentration such as sodium cholate, octyl glucoside, or Chaps (cholamidopropyl-dimethylamonio-propane sulfonate) works best.

A basic outline of the reconstitution procedure is as follows. First, purify the individual protein components (for examples of protein purification, see ref. 10 for stimulatory G-protein (G_s) and ref. 11 for β-adrenergic receptor). Combine these proteins with lipids, either by first reconstituting one protein, and adding the others to these vesicles; or by adding all of the proteins to detergent-solubilized lipid at once. Next, the detergent which solubilized the purified proteins and lipids must be removed. If the detergent was cholate or octyl-glucoside, it may be possible simply to dilute the remaining detergent by the addition of a large volume of buffer. For most detergents, including cholate and octyl-glucoside, gel filtration is often necessary (for instance, using a column with material such as Sephadex G-50). Vesicle-containing fractions may be detected either by turbidity (they will appear cloudy due to the presence of lipid) or by a protein activity assay, such as radioligand-binding. *Protocol 3* is an example of a reconstitution protocol, for the β-adrenergic receptor and G_s. If this protocol is not successful, altering the original protein purification, lipid(s) used in reconstitution, and/or the method of detergent removal will be necessary (for an overall review of reconstitution methods, see ref. 12).

Protocol 3. Reconstitution of the ß-adrenergic receptor and G_s

You will need:

- buffer 1 (10 mM Hepes, pH 8.0, 1 mM EDTA, 0.1 mM DTT (dithiothreitol), 0.1% Lubrol)
- buffer 2 (20 mM Hepes, pH 8.0, 1 mM EDTA, 0.1 mM NaCl, 2 mM $MgCl_2$, 1 mM DTT)
- buffer 3 (20 mM Hepes, pH 8.0, 1 mM EDTA, 0.05% digitonin)
- lipid (0.4 mg dimyristoylphosphatidylcholine, 0.8 mg turkey erythrocyte polar lipids, 3.6 mg deoxycholate, 0.4 mg cholate in 1 ml buffer 2)
- purified β-adrenergic receptor (0.2–0.7 mg/ml) in buffer 3
- purified G_s (1–10 mg/ml) in buffer 1
- 12.5% (w/v) sucrose solution (12.5 g sucrose, water to 100 ml total volume)

Protocol 3. *Continued*

• 50% (w/v) sucrose solution (50 g sucrose, water to 100 ml total volume)

1. Do entire reconstitution at 0–4 °C.

2. Prepare Sephadex G-50 column. Volume of column should be at least 10× that of sample volume.

3. Equilibrate column with buffer 2.

4. Mix 100 μl receptor, 60 μl G_s, and 120 μl lipid.

5. Load onto Sephadex G-50 column.

6. Elute vesicles from column with buffer 2.

7. Collect turbid fractions with high β-adrenergic agonist-binding activity.

8. Place 4 ml of 50% sucrose on the bottom of an ultracentrifuge tube suitable for use with a SW41 rotor. Carefully layer 4 ml of 12.5% sucrose on top. Layer 3 ml of turbid fraction/tube on to these gradients, without disturbing the gradient. Centrifuge for 4 h at 280 000 *g* in a SW41 rotor.

9. Collect material at interface between layers of sucrose (fraction should appear cloudy).

10. Freeze vesicles and store at −80 °C until used.

---

3. Use of agonist-binding to detect receptor/ G-protein interactions

It is difficult to detect protein conformational changes in complex systems (for example, receptors and G-proteins). In order to study the behaviour of these scarce proteins in a relatively unperturbed membrane environment, indirect methods such as radioligand-binding or enzymatic activity measurements are often employed. The basic principle underlying the use of agonist-binding to detect receptor G-protein interactions is that any parameter of the system (for example, agonist affinity) which changes with the conformation of the receptor may be used as a monitor of RG coupling. It has been known for years that the binding affinity of agonists such as glucagon (13) and isoproterenol (7) is reduced by GTP. In general, antagonist-binding is not. Indeed, these two cases provided the first examples of how direct ([³H]glucagon) and indirect (isoproterenol competing for [¹²⁵I]IHYP (iodohydroxypindolol)) binding methods can be used to detect RG coupling. Such data have lead to the concept that receptors exist in two main conformational states, R (receptor alone) and RG (receptor coupled to the G-protein). The former has a low affinity for agonists while the latter has a high affinity. This section will explore uses of agonist-binding in the detection of RG coupling and evaluate the advantages and disadvantages of direct vs. indirect methods.

3.1 Competition methods: GTP shifts

3.1.1 Background

Radioligand-binding studies often employ labelled antagonists. *Table 2* illustrates the properties of agonist and antagonist radioligands. Previously clear-cut areas have become fuzzier as more aspects of RG interaction are explored. For example, antagonists were previously thought to bind to receptor without regard for the G-protein, but some antagonists bind to the RG complex with *lower* affinity than to free R. In this situation they may actually inactivate a basal level of G-protein function (14, 15).

Table 2. Properties of antagonist and agonist radioligands

Radioligand	Properties	Uses
Antagonist	High-affinity Low dissociation rate-constant Insensitive to conformational state	Determining total receptor numbers Competition experiments with unlabelled compounds
Agonist	High and low affinity binding Sensitive to conformational state	Identifying RG coupling

To assess RG coupling, radiolabelled antagonists must be used indirectly. Competition by the β-adrenergic agonist isoproterenol for $[^{125}I]IHYP$ binding (7) showed that GTP could greatly decrease the agonist's affinity for the β-adrenergic receptor. These 'GTP shifts' have been used as evidence that other receptors interact with a G-protein such as the α_1-adrenergic receptor (16). Non-hydrolysable GTP analogues such as 5'-guanylylimidodiphosphate (GppNHp) and guanosine 5'-O-3-thiotriphosphate (GTPγS) are most frequently used because they are not degraded by the intrinsic GTPase activity of the G-protein.

3.1.2 Experimental considerations

In order to detect high-affinity agonist-binding in binding assays (either direct or competition), it is necessary to eliminate contamination of the membrane preparation by endogenous GTP or GDP (see *Table 1* and Section 2.1). If agonist competition experiments do not reveal a biphasic curve or if there is no 'GTP shift', then:

- Prepare membranes with reduced guanine nucleotide concentrations (see *Protocol 1* above).

- Increase the volume of the binding assay to dilute any endogenous nucleotides.

- Try longer equilibration times (60–180 min) to see the full appearance of the high-affinity component of agonist-binding.

For example, epinephrine competition curves for [³H]yohimbine binding to the platelet α_2-adrenergic receptor show only a small GTP shift when performed with approx. 100 μg of protein in 100 μl of buffer, but a much more prominent GTP shift is seen in 500 or 1000 μl.

3.1.3 Data analysis

The curves that result from agonist competition experiments (see *Figure 1*) are usually complex (i.e. have a Hill slope of less than 1). Such data are generally best analysed by a non-linear least-squares method (17, 18). The readily available computer software for such analysis usually includes models with a single affinity site with a variable Hill slope, or multiple independent binding sites. In designing an experiment for analysis by a two-site model, several considerations should be kept in mind:

- Space data points evenly on a logarithmic scale (for example, 1, 3, 10, 30, etc., or 1, 2, 5, 10, 20, 50, etc.)

- Ten to 12 concentrations of competitor are usually the *minimum* for adequate analysis by a two-site model. Each concentration is replicated in two to four tubes.

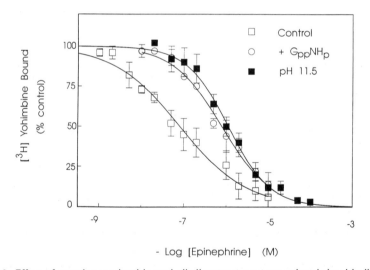

- Log [Epinephrine]   (M)

Figure 1. Effect of guanine nucleotide and alkaline treatment on epinephrine binding to human platelet membranes. Platelet membranes were pre-treated for 1 h with 50 mM sodium phosphate buffer at pH 7.6 (□, ○) or pH 11.5 (■) and the membranes pelleted and resuspended in buffer (50 mM Tris, 10 mM MgCl₂, 1 mM EGTA, pH 7.6). Binding of 10 nM [³H]yohimbine was measured in the presence of the indicated concentrations of epinephrine with (○) or without (□, ■) 10 μM GppNHP. The solid lines are curves fitted to a Hill equation. Modified from ref. 25, with permission.

10

● A saturating concentration of the unlabelled ligand should be included in a set of tubes to define non-specific binding.

The effect of guanine nucleotides on such agonist competition curves is to decrease the affinity and increase the Hill slope (*Figure 1*). Alkaline treatment of membranes, an alternative method for inactivating G-protein (19, 20), also decreases agonist binding and increases Hill slope (*Figure 1*). In a two-site model, nucleotides decrease the affinity of the agonist at the high-affinity site or decrease the proportion of high-affinity sites.

Protocol 4. Non-linear curve fitting for competition data

1. Plot all raw data as bound c.p.m. (or fmol/mg protein) vs. log concentration of competitor.

2. Check that replicates agree within 5–10%. Obvious outliers may be dropped to prevent skewing of fitted parameters. If there are more than one or two such bad points it is probably necessary to repeat the experiment.

3. Using a non-linear least-squares regression program (this example will use GraphPAD InPlot), enter the data as bound radioligand vs. log [competitor]. Be sure to include the data for zero competitor.

4. Estimate the concentration of competitor that reduces radioligand binding by 50%.

5. Using this concentration (or the default computer-generated estimate if your program provides one) as the initial estimate of IC_{50}, fit the data to the one-site competition model.[a] All parameters, maximum binding, minimum binding, and IC_{50}, should be allowed to vary.

6. Record the sum of the squared deviations of the data from the fitted equation. This value (SS_1) and the degrees of freedom (df_1 = number of data points − number of free parameters in the one site fit)[b] will be needed to statistically test fits produced by the different models.

7. Plot the original data and the results of the fitted equation. It is essential to examine the predictions of the fit to be sure that the fit adequately reflects the data.

8. If the predicted curve is too steep or too shallow to accurately describe the experimental data, then refit the data with a Hill model.[c] For G-protein-coupled receptors the Hill slope for agonist competition is usually less than 1.0.

9. The IC_{50} is a descriptive variable that relates the effective concentration under the experimental conditions. The researcher is usually interested in the affinity of the ligand for the receptor (i.e. the K_i). This can be estimated from IC_{50} by the Cheng–Prusoff relation:

Protocol 4. *Continued*

$$K_i = IC_{50}/(1+[D^*]/K^*_d)$$

where $[D^*]$ and K^*_d are the concentration and dissociation constant for the radioligand.

[a] The equation for a one-site competition model is:

$$B = \min + \frac{(\max-\min) * IC_{50}}{1 + IC_{50}}$$

where B is the amount bound, I is the concentration of the competing ligand, max is the binding with no competitor, min is the binding with saturating concentrations of competitor (which should equal the non-specific binding) and IC_{50} is the concentration of competitor causing half-maximal competition.

[b] There are three free parameters in the one-site model: maximum binding, minimum binding (i.e. non-specific), and the IC_{50}.

[c] The equation for the Hill model is:

$$B = \min + \frac{(\max-\min) * IC_{50}{}^{nH}}{I^{nH} + IC_{50}{}^{nH}}$$

where nH is the Hill slope and the other parameters are as in footnote [a]. Some authors call nH a 'pseudo' Hill slope when used in competition curves, but the simpler nomenclature of Hill slope will be used here.

In a few cases, the binding of antagonists to a G-protein-coupled receptor is *increased* by addition of GTP. This would occur if binding of the antagonist to the RG complex were weaker than to R alone *and* the receptor and G-protein existed in a pre-coupled complex. An increase in [³H]spiperone binding to the pituitary dopamine D_2 receptor has been invoked as evidence that D_2 receptor is pre-coupled to a G-protein in the absence of ligands (14). Kinetic modelling studies of the α_2-adrenergic receptor (21) and GTPase measurements with the δ-opiate receptor (22) have confirmed the existence of pre-coupled RG complexes (see Section 4.3.2). This has interesting physiological implications but also has an impact on data analysis. Calculation of the K_i for a competing ligand requires knowledge of the K_d^* of the radioligand, so to fully assess the effect of guanine nucleotides (or any other treatment) on the K_i of a competitor, the K_d^* of the radioligand must also be determined under all experimental conditions (for example, with and without GTP).

Protocol 5. Fitting a two-site model

1. Enter the data into your non-linear least-squares regression program and analyse by the one-site competition model (*Protocol 4*).

2. Estimate the initial parameters of the two-site model.[a]

   (a) Identify the maximum and minimum values for radioligand-binding.

   (b) If the competition curve has two obvious phases, estimate the percentage of binding in the high- and low-affinity phases; otherwise use 50% as a starting estimate.

12

(c) Estimate the IC_{50} for the high- and low-affinity components or pick values approximately three- to tenfold above and below the IC_{50} for the one-site fit.

3. Plot the curve generated by the initial estimates and compare it to the raw data. If there is a great discrepancy between the experimental and theoretical results, you may wish to enter new estimates of the parameters. How close the initial estimates must be for the non-linear regression program to converge to an acceptable fit depends on the quality of the data and the specific computer program you use.

4. Run the program to estimate the binding parameters for the two-site model.

5. If the fitting procedure does not converge to reasonable parameter estimates (i.e. no estimate or physically unreasonable ones such as negative values for bound ligand or IC_{50}s) then:

   (a) Repeat step 3 to manually adjust the initial parameters to give the program a better initial estimate.

   (b) Examine the data for the presence of outliers that may prevent adequate fitting.

   (c) Try constraining the minimum or maximum binding to the non-specific (minimum) and/or control (maximum) binding.

   (d) Consider using a different weighting scheme to take into account the different variances of the binding at different concentrations of competitor (see refs 17 and 18).

6. Record the binding parameters, the sum of the squared residuals (SS_2) and degrees of freedom (df_2) for the two-site fit.

7. To determine if the two-site model is statistically better than the simpler one-site model, use the variance ratio test (F-test, refs 17 and 18).

   (a) Calculate the F statistic from the SS and df values from step 6 of *Protocol 4* and step 6 of this protocol according to the following equation.

$$F = \frac{(SS_1 - SS_2)/(df_1 - df_2)}{(SS_2/df_2)}.$$

   (b) Look up the associated probability (P) value from F using ($df_1 - df_2$) and df_2 as the degrees of freedom in the F table.

   (c) A small P value (for example < 0.05) indicates that the more complex two-site fit provides a statistical improvement greater than that expected from just introducing additional free parameters.

---

[a] Your computer program may provide initial estimates so that you can skip this step (GraphPAD Inplot does provide reasonable estimates).

While it must be recognized that these types of analyses do not provide parameters of a physical model such as the ternary complex model (14, 23), both the Hill model and the two-site model provide useful phenomenological descriptions of the data. For example, if a guanine nucleotide produces an increase in the K_i for an agonist in competition experiments accompanied by an increase in the Hill slope, that would be evidence that the receptor interacted with a G-protein. To statistically analyse the significance of the shifts an F test can be used again (*Protocol 5*). More mechanistic conclusions would require analysis with a specific model (Section 4).

Protocol 6. Testing for significant shifts in competition curves

1. Enter the control (c) and experimental (e) data sets into the non-linear least-squares program (for example, GraphPAD InPlot).

2. Create a third data set with the control and experimental data combined (ce).

3. Fit the combined data set to the Hill model (*Protocol 4*) to generate SS_{ce} and df_{ce}. Record these values as well as max and min (see footnote *c* in *Protocol 4*).

4. Fit the control and experimental data sets to generate SS_c, df_c, SS_e, and df_e, respectively. For this step, constrain (or fix) the min and max to the values from step 3.[a]

5. Calculate the F statistic from this equation:

$$F = \frac{(SS_{ce} - [SS_c + SS_e])/(df_{ce} - [df_c + df_e])}{([SS_c + SS_e]/[df_c + df_e])}.$$

6. Look up the P value from the F table using $df_{ce} - [df_c + df_e]$ and $[df_c + df_e]$ as the degrees of freedom.[b]

7. A significant P value means that the fit got significantly worse when the data sets were combined. The interpretation of this would be that the results of the experimental and control measurements are different from each other.

[a] This will make the statistical question simply 'Are there differences in the IC_{50} or nH between the control and experimental data?'. This question can be refined to ask 'Is there a difference in the Hill slope between control and experimental?' by fitting the data with min, max, and Hill slope constrained to the value for the combined fit, then fitting again with the Hill slope unconstrained.

[b] If the procedure is followed exactly, the value of $df_{ce} - [df_c + df_e]$ should equal 2. This is the number of variables that are constrained in the combined fit but allowed to float in the separate fits (i.e. IC_{50} and nH).

As an example of this type of analysis we will examine (*Table 3*) the SS and df values and the calculation of *F* for the data from *Figure 1*.

Table 3. Statistical analysis of GppNHp effect on α_2-adrenergic agonist-binding

Data set	SS	df
Control	3140	54
+ GppNHP	1323	30
Combined (Control and + GppNHp)	16715	88

From these values we can calculate:

$$F = \frac{(16715-[3140+1323])/4}{[3140+1323]/84} = 57.7.$$

The statistical significance is evaluated by looking up an *F* of 57.7 with 4 and 84 degrees of freedom. The *P* value can be found in the table of *F* values in a standard statistics textbook. (Two values are needed for the degrees of freedom: the value for the numerator is the difference in the number of free parameters between the individual and combined fits, and the denominator is the sum of the degrees of freedom for the individual fits.) The larger the value of F, the more significant the difference. In this case, the difference between the control and GppNHp conditions is highly significant ($P < 0.00001$). The value of *F* for control versus pH 11.5 is 126, also with a *P* value of < 0.00001. A comparison of GppNHp with pH 11.5 gives a much lower *F* of 5.4 since the shift in the competition curve is much smaller. Because of the large number of data points (93 degrees of freedom) the difference does reach statistical significance ($P = 0.0006$) despite the small shift in the curve.

3.2 Direct agonist-binding: equilibrium

3.2.1 Background

Since agonist-binding is highly dependent on the receptor conformational state, direct measurements of radiolabelled agonist binding are easily perturbed by environmental variables. This is a disadvantage if one simply wants to quantitate the number of receptors but can be very useful for studying factors that affect RG coupling. The difference in affinity of the agonist for the high and low affinity conformations of the receptor may be 100-fold or more. In this case, direct agonist binding measurements over a limited concentration range will detect only the high-affinity state. This is well-illustrated for the α_2-adrenergic receptor (*Figure 2*). The antagonist [³H]yohimbine binds to the full complement of α_2-receptors while the full agonist [³H]UK-14 304 and partial agonist [¹²⁵I]*p*-iodoclonidine only detect about 40% of the [³H]yohimbine sites. These agonists do bind to all of the [³H]yohimbine sites as shown by competition experiments (*Figure 1*) but their affinity is so low at 60% of the sites that they are not detected in direct agonist-binding measurements.

3.2.2 Experimental considerations

As in competition assays, it is important to keep the membrane protein concentration low to minimize endogenous GTP contamination. This can be done by using sucrose gradient purified plasma membranes which have less guanine nucleotide contamination (4, 7) or by increasing the volume of the assay.

Protocol 7. Equilibrium agonist-binding to platelet α_2-adrenergic receptors

You will need:

- purified platelet plasma membranes (approx. 1 mg/ml protein in buffer 1)
- [³H]UK-14 304 (bromoxidine; α_2-adrenergic agonist)
- non-radioactive yohimbine [10^{-4} M] (α_2-adrenergic antagonist)
- buffer 1 (50 mM Tris, pH 7.6, 10 mM $MgCl_2$, 1 mM EGTA)
- buffer 2 (50 mM Tris, pH 7.6, 10 mM $MgCl_2$)

1. Prepare 2 ml of a 50 nM stock solution of [³H]UK-14 304. Place the following volumes into tubes to obtain the indicated final concentrations: 200 µl (20 nM), 150 µl (15 nM), 100 µl (10 nM), 50 µl (5 nM), 20 µl (2 nM), 15 µl (1.5 nM), 10 µl (1 nM), 5 µl (0.5 nM). For an experiment with triplicates for both specific and non-specific points, six tubes will have each of the above concentrations.

2. For the non-specific points, add 50 µl of non-radioactive yohimbine to each tube. To all other tubes, add 50 µl of buffer. Add sufficient buffer 1 for a final volume of 400 µl.

3. Initiate the incubation by adding 100 µl of platelet membrane (0.1 mg protein). Incubate at room temperature for 60–90 min.

4. End incubation by adding 5 ml of ice-cold buffer 2 to each tube and then filter immediately. Wash filters twice with 10 ml of ice-cold buffer 2. Use Whatman GF/C filters or equivalent.

5. Allow filters to dry completely (30–60 min), and then place each filter in a scintillation vial. Add 4.5 ml of scintillation fluid (such as Scintiverse) to each vial. Cap tightly and place in scintillation counter.

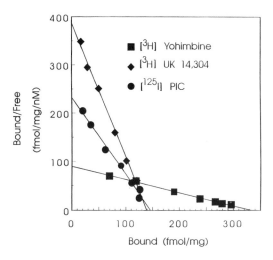

Figure 2. Binding of agonists and antagonists to α_2-adrenergic receptors in human platelet membranes. Saturation binding curves for the antagonist [³H]yohimbine, the partial agonist [¹²⁵I]p-iodoclonidine and the full agonist [³H]UK–14 304 were measured. A Scatchard plot of the binding is shown. Reproduced from ref. 8, with permission.

3.2.3 Data analysis

When low concentrations of the agonist are used and the binding isotherm is hyperbolic, only the high-affinity RG state of the receptor is measured. In this case, data analysis is simple and may be done by linear regression of a Scatchard transformation of the data. A more statistically valid method even with linear Scatchard plots, is analysis of the raw binding data by non-linear regression analysis (see *Protocols 4* and *5*). When high concentrations of the radiolabelled agonist are used, complex saturation binding isotherms may be observed. This is usually due to binding to the low-affinity as well as the high-affinity forms of the receptor. In this case a non-linear regression analysis is essential to obtain accurate estimates of the binding parameters. One must rule out experimental artefacts that can result in curvilinear Scatchard plots (see ref. 24). To demonstrate that the high-affinity binding is to a G-protein-coupled state of the receptor, measure agonist binding in the presence of a guanine nucleotide analogue. For α_2-adrenergic, M4 muscarinic and β-adrenergic receptors, GTP analogues reduce high-affinity agonist binding by 80–90%, indicating that most of the binding corresponds to an RG complex.

3.3 Advantages of direct agonist-binding in detection of RG coupling

Two major advantages of the direct measurement of agonist-binding for assessing RG interactions are:

- the signal-to-noise ratio for changes in agonist-binding is very good (as compared with competition methods), and
- it can be easily studied on very rapid time-scales.

In human platelet membranes, high-affinity α_2-adrenergic agonist-binding is completely eliminated by treatment at pH 11.5 which inactivates the inhibitory G-protein (G_i, ref. 19). The reconstitution of [^{3}H]UK-14 304 binding by incorporation of purified 'other' G-protein (G_o, ref. 25) shows the excellent signal that is achieved with direct agonist-binding (*Figure 3a*). There is a 10 to 20-fold increase in agonist-binding upon addition of G_o. Also, the increase in agonist-binding can be monitored at a single radiolabelled agonist concentration as opposed to the full competition curves that are usually required for the indirect methods (*Figure 3b*).

3.4 Direct agonist-binding: kinetics

3.4.1 Background

The difference between direct agonist-binding and indirect competition experiments becomes more apparent in kinetic studies. With a radiolabelled agonist it is easy to measure binding on a time-scale of seconds. The radioligand and receptor are mixed and samples taken for direct measurements of binding by standard filtration methods. The early kinetics are often lost using competition methods because of the slow binding of the antagonist radioligand. The methods for making kinetic measurements will be described in this section and their use in the analysis of RG coupling described in Section 4 on modelling.

3.4.2 Experimental considerations

The materials and basic aspects of kinetic methods are identical to those for equilibrium binding (*Protocol 7*), but the reaction is carried out in a single flask or tube with a sample taken at each time point. This improves reproducibility and greatly simplifies the experiment. For any kinetic study on a different system, the time to reach equilibrium (3600 sec in *Protocol 8*) must be determined in pilot studies on the minute (rather than second) time-scale.

Protocol 8. Kinetic binding measurements

You will need:

- buffer 1 (50 mM Tris, pH 7.6, 10 mM $MgCl_2$, 1 mM EGTA
- buffer 2 (50 mM Tris, pH 7.6, 10 mM $MgCl_2$)

1. Prepare a stock solution[a] of the radioligand (for example, 13 ml of 2 nM [^{3}H]UK-14 304) in buffer 1. This represents a 20% excess to ensure an adequate amount of sample for the last (equilibrium binding) measurements.

18

2. Prepare an equal volume of diluted platelet membranes (0.2 mg protein/ml) in buffer 1.

3. Pre-warm radioligand and membranes 2–3 min in a shaking water bath to reach the experimental temperature (for example, 25 or 37 °C).

4. Rapidly mix equal volumes of the ligand and membrane and return to temperature bath.

5. Remove 1 ml samples at appropriate times (see above). Take single samples for times up to 90 sec and duplicates (done 5 sec before and 5 sec after the recorded time) for 120 sec on.

6. Immediately, dilute in 5 ml of iced buffer 3, filter and wash 2 × with 10 ml of iced buffer 3. This should take no longer than 10 sec. For the rapid early time points it is often helpful to have two people doing the experiment. One takes the aliquot and dilutes it, and the other filters the sample and washes the filter.

7. Repeat for the non-specific binding measurement with 6.5 ml of [³H]UK-14 304 and membranes to which a saturating concentration of competing ligand (10 µM oxymetazoline) has been added approx. 30 min before.[b]

8. Plot total and non-specific binding as a function of time and calculate specific binding (see footnote to *Table 4*).

9. Fit the data for specific binding vs. time to an exponential association function: $B = B_\infty (1-e^{-kt})$, where B equals bound ligand at time t, B_∞ is bound ligand at equilibrium (i.e. 3600 sec) and k is the rate-constant for the binding reaction. The non-linear least-squares regression routines in GraphPAD InPlot may be used for this purpose. Alternatively, a semilogarithmic plot of ln $(B_\infty/(B_\infty - B))$ vs. t may be prepared. If the kinetics follow a simple exponential association, a straight line should result with y-intercept zero and slope k.

[a] Note: the ligand concentration is twice the final concentration desired.
[b] The non-specific samples may be interspersed with the total binding samples if the times are carefully planned out in advance. If so, be sure to change pipette tips between non-specific samples and total samples because just a few microlitres of the non-specific buffer may contain enough competing ligand to cause dissociation of the radioligand in total samples.

Some potential difficulties in kinetics studies and their solutions are outlined in *Table 4*.

For a simple bimolecular binding reaction, the kinetics of both ligand association and dissociation are exponential. The theory of such binding is well-described in standard references (24, 26). The time-dependence of high-affinity α_2-adrenergic agonist binding may be used as a measure of the time-course of RG coupling (21, 27; see Section 4.3.2 regarding modelling of conformational changes).

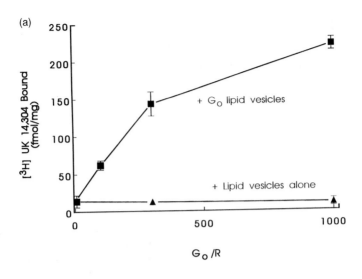

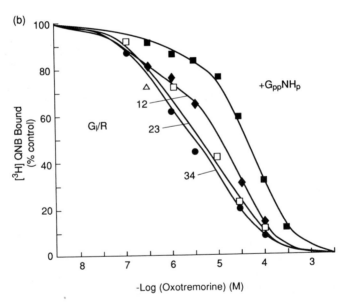

Figure 3. Reconstitution of α_2-adrenergic and muscarinic agonist-binding with purified G-proteins. (a) Binding of the full α_2-adrenergic agonist [^{3}H]UK–14 304 (1 nM) to alkaline-treated human platelet membranes is illustrated. The effect of reconstituting with purified bovine brain G_o or control lipid vesicles is shown. (Modified from ref. 25, with permission). (b) Competition assay of binding of the agonist oxotremorine to purified muscarinic receptors reconstituted with purified bovine brain G_i is shown. The numbers on the figure represent the molar ratio of G_i to receptors. (Modified from ref. 54, with permission.)

20

Table 4. Experimental considerations in binding kinetics

Difficulties	Solutions
Inadequate mixing resulting in high local concentrations of ligand or membrane	Use equal volumes of ligand and receptor.
	Use a mixing method rapid enough for the time-scale of the experiment
	Minutes: hand mixing
	Seconds: syringe mixer.
	Test mixing system by use of coloured dyes before use in experiments.
Time-dependent non-specific binding	Always include parallel kinetic measurements of non-specific binding.[a]
Temperature-dependent binding	Conduct incubation in a temperature-controlled bath.
	Pre-warm reagents to desired temperature before initiating reaction.
Rapid association and dissociation	Use separation method fast enough (for example, filtration) to permit sampling in real time.
	Use separation method fast enough to avoid ligand dissociation during separation.

[a] Non-specific binding is usually much less time-dependent than specific binding so fewer non-specific time points may be collected. For example, if total binding was measured at 10, 20, 30, 40, 50, 60, 75, 90, 120, 150, 180, 240, 300, 600, and 3600 sec, non-specific could be measured at 10, 30, 60, 120, 300, 600, and 3600 sec. If non-specific binding were found to be constant or only varying slightly, the non-specific binding could be approximated by a linear regression of the measured data over a range of times (for example, 10–600 sec).

4. Modelling

4.1 Introduction

Because RG interactions take place in a membrane environment, conformational changes of the proteins involved must often be assessed indirectly. In order to determine the molecular constants of RG interactions from indirect measurements, such as agonist-binding, enzymatic activities, and even spectroscopic measurements, we must have an explicit model of the interactions. The predictions of the model are then compared with the data to: (a) choose between alternative models, and (b) estimate molecular rate and binding constants.

4.2 Steps in testing biological models

Each type of experimental system presents unique aspects when one develops mathematical models. Despite that, there are certain steps used for any type of modelling. A general approach is outlined here.

(a) Write down the model with equilibrium or rate-constants.

(b) Pick the variable (concentration of a component or rate of an enzymatic activity) that will be observed experimentally.

(c) Write the equations that govern the behaviour of the model(s).

(d) Solve the equations for the observable variable.

  i. Try to express the observable variable in an equation (analytical solution)

  or

  ii. Use a numerical simulation method to approximate the solution.

(e) Predict the value of the observable variable under a variety of conditions.

(f) Choose conditions that give very different predictions of the observable variable for the possible models.

(g) Experimentally test the predictions.

(h) Revise the model and start again.

The first two steps force the investigator to prepare an explicit description of the model that can be evaluated mathematically. Steps (c) to (e) can be done by a mathematician without regard to the biology of the system. The last steps require a collaborative effort between the experimentalist and the theorist/mathematician.

4.3 Ternary complex model

The ternary complex model of De Lean *et al.* (28) provides a good starting point for examining RG interactions *Figure 4* illustrates two versions of the ternary complex model. Model 1 could be considered an induced-fit model where drug binding induces RG coupling, while Model 2 is the cyclic ternary complex model which allows RG interaction before ligand binds (i.e. pre-coupled RG complex). K_1 is the affinity constant for drug binding to R alone, K_3 is its affinity for the RG complex, K_4 is the affinity of G for R alone, and K_2 is the affinity of G for the DR complex. In the cyclic model, the four affinity constants must follow the relation $K_1 \cdot K_2 = K_3 \cdot K_4$. Thus, if drug binds more tightly to RG than to R then G must bind more tightly to DR than to R.

4.3.1 Equilibrium agonist-binding

We and others examined the predictions of the ternary complex for equilibrium agonist-binding (14, 28, 29). The observable variable in the models is agonist-binding which is (DR + DRG). Using either Model 1 or Model 2, the ternary complex model only accounts for the complex (i.e. two-site) equilibrium agonist-binding if there is a nearly stoichiometric amount of G with respect to R. This hypothesis was tested by measuring the amount of

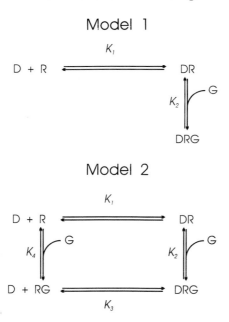

Model 1

Model 2

Figure 4. Models of receptor G-protein coupling. Two versions of the ternary complex model of ligand-binding to G-protein-coupled receptors are shown. Model 1 only permits G-protein coupling to the receptor *after* ligand has bound to the receptor. Model 2 includes interactions of the receptor and G-protein both before and after ligand-binding.

pertussis toxin substrate in platelet membranes (29) and a 20- to 100-fold excess of G_i over α_2-adrenergic receptor was found. Similar ratios have been seen for β-adrenergic receptors and G_s in S49 cells (30). Since the prediction of the model (that there should be equal amounts of G and R) did not hold, a modification of the ternary complex model was necessary. The existence of a form of the receptor, R', that could never couple to G was postulated (21, 29). This modified ternary complex model was then evaluated with agonist-binding kinetics to distinguish between the induced-fit and the pre-coupled models.

4.3.2 Agonist-binding kinetics

The concept of using agonist-binding kinetics as a monitor of receptor conformational state derived from the work of Boyd and Cohen (31, 32) on the nicotinic acetylcholine receptor. In that system the conformational change of the receptor was an intramolecular event and did not require interaction with a G-protein. The same methods and analysis are easily applied to the G-protein-coupled systems.

The main difference between the induced-fit ternary complex model and the cyclic ternary complex model is the existence of a pre-coupled RG complex in Model 2 which would predict that high-affinity GTP-sensitive binding should appear quite rapidly. Thus, we studied the kinetics of α_2-adrenergic agonist-binding to assess the state of RG coupling at these early times (21). The steps involved in this analysis will be outlined to illustrate an approach to modelling complex biological systems.

(a) *Write down the model with equilibrium or rate-constants.*

Figure 5 shows the actual model evaluated. It is the same as the cyclic ternary complex model but replaces each equilibrium constant with a pair of rate-constants and includes the uncoupled receptor, R′. This model is quite complex with 10 free parameters.

(b) *Pick the variable (concentration of a component or rate of an enzymatic activity) that will be observed experimentally.*

As described for the equilibrium ternary complex models, the observable variable in agonist binding kinetic studies is total agonist-binding (DR′ + DR + DRG). Thus, experimentally, we cannot simply distinguish among agonist bound to the three forms of the receptor. However, the ligand concentration-dependence and time-course of binding and dissociation will be different for the three, so the kinetics will contain information about all three receptor forms.

(c) *Write the equations that govern the behaviour of the model(s).*

(d) *Solve the equations for the observable variable.*

(e) *Predict the value of the observable variable under a variety of conditions.*

The differential equations for this (and three simpler models) were written down but they were too complex to solve exactly (21). The kinetic simulation computer program SAAM was used to simulate the binding kinetics (33). The time-course of binding was predicted (and observed) to have fast and slow kinetic components. The rate and amount of the two components differed significantly among the four models tested and two could be excluded. Four models were tested in ref. 21. The model in *Figure 5* was simplified by leaving out different components. When R and DR were eliminated (i.e. R′ and RG are not interconvertible), it became a 'two-receptor' model with no real conformational change. When RG was left out it became Model 1, plus an uncoupled form of the receptor. When DR was left out it became what we called a 'back door' model in which the appearance of DRG only occurred through the pre-coupled state and not via an induced fit. Neither the 'two-receptor' nor the 'back door' model could account for the concentration-dependence of the binding kinetics.

Models 1 and 2 (with an additional uncoupled receptor, R′) remained after this initial analysis, and the parameters were optimized for the two models. The optimization was done by simulating the binding kinetics at six different ligand concentrations with different sets of parameters to minimize the sum of the squared residuals between the experimental data and the theoretical fit.

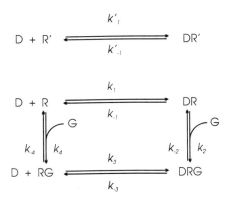

Figure 5. Kinetic model of α_2-adrenergic receptor/G_i coupling. A modified ternary complex model was used by Neubig *et al.* (21) to analyse the time-course of α_2-adrenergic agonist-binding. R' is a form of the receptor that can never couple to G_i. R is not coupled but is in equilibrium with the RG complex. The *k* values represent rate-constants for the individual steps in the interaction of receptor and G-protein (see text and ref. 21 for details).

To do this we made use of both manual adjustments of the parameters and the automatic optimization capabilities of SAAM. Since there were many parameters, we utilized additional measurements to constrain some of them. For example, k_{-3} governs the slow dissociation of ligand from the DRG complex. This was measured in simple dissociation kinetics experiments, so k_{-3} was fixed at the measured value of 4.3×10^{-4} sec (corresponding to a half-time of 27 min) while the other parameters were optimized.

(f) *Choose conditions that give very different predictions of the observable variable for the possible models.*

Model 2 predicts that there will be a pre-coupled RG complex resulting in rapid appearance of a ternary complex, DRG, while Model 1 predicts that all of the fast binding would be to the lower affinity forms DR or DR'. As we have seen (Section 3), the ternary complex, DRG, can be identified by its GTP sensitivity. We simulated the effect of adding GTP at various times during the measurement of agonist-binding kinetics. At 5 min, a time when almost all of the binding was due to the fast kinetic component, Model 2 predicted (solid lines in *Figure 6*) that much of the agonist bound would dissociate rapidly in the presence of GTP. Since there is no pre-coupled RG complex in Model 1, it predicted very little decline in binding upon addition of GTP (dashed lines in *Figure 6*).

(g) *Experimentally test the predictions.*

An experiment was done using the conditions of the simulation. Model 2 could account for the binding kinetics and effects of GTP quite well. This experiment ruled out Model 1 and provided good evidence for Model 2. One

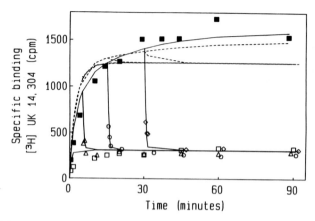

Figure 6. GTP-dependence of early α_2-adrenergic agonist binding. The time-course of binding of [^{3}H]UK–14 304, a full α_2-adrenergic agonist, was measured, and GppNHp added at 5, 15, and 30 minutes after initiation of binding. Theoretical fits of the time-course are shown for Model 1 (*dashed lines*) and Model 2 (*solid lines*). The rapid appearance of GTP-sensitive agonist-binding and the excellent fit by Model 2, provide evidence for the existence of a pre-coupled α_2-adrenergic receptor–G$_i$ complex. Experimental details can be found in ref. 21.

feature to note about this test is that the parameters were optimized for one set of data, the concentration-dependence of the binding kinetics, while the test examined a completely different aspect, the effect of GTP. Thus, the model did a very good job of predicting behaviour that was not explicitly included in the parameter optimization. This is a key feature of testing models. A robust model should predict behaviour not explicitly built into the model.

(h) *Revise the model and start again.*

In this case, the fits with Model 2 were good enough and the biological system complex enough that we did not attempt to further refine the model simply for binding studies. Instead, we chose to examine additional predictions of a model that included function of the G-protein (see Section 5.2).

5. Functional methods

5.1 GTP binding and GTPase activity

There are a number of functional assays which can indirectly assess the RG interaction. These include GTPase assays, which measure the ability of the G-protein to hydrolyse GTP; GTPγS binding, which measures the number of G-proteins binding guanine nucleotides; and second messenger assays, which measure the stimulation of adenylyl cyclase, phospholipases, or other G-protein-coupled effector systems. Details of these second messenger methods are described in Chapters 2–8 in this book.

Ross and co-workers have developed detailed models of the mechanism of β-adrenergic receptor-mediated G_s activation. They measured GTPase activity and guanine nucleotide binding and release with β-adrenergic receptor and G_s reconstituted in lipid vesicles (34, 35). While these studies have yielded important information about RG mechanisms, the interaction between receptor and G-protein was treated as a simple catalytic process without including the reciprocal interactions of receptor and G-protein conformational changes. Tota and Schimerlik (36) utilized a similar reconstitution system to examine muscarinic receptor–G_i mechanisms. By studying the dependence of GTPase activity on receptor concentration, they were able to estimate the affinity of receptor for G-protein with either a full agonist or a partial agonist bound to the receptor. They concluded that a full agonist resulted in a higher affinity of the muscarinic receptor for G_i than a partial agonist. Interestingly, the V_{max} for GTPase activity was the same with the full and partial agonist.

5.2 Second messenger measurements

Functional assays are important for examining RG interactions because they permit dynamic measurements of the system, and particularly of the G-protein. Since most ligand binding to receptor is measured in the absence of guanine nucleotides, the system could be considered a frozen snapshot of the cell membrane. Guanine nucleotides are present during functional assays, allowing the RG cycle to be completed. Information from ligand binding and functional assays is complementary and can be used to derive more complete models of the system.

Thomsen *et al.* (5) combined an explicit model of α_2-adrenergic receptor and G_i conformations with measurements of α_2-adrenergic agonist-binding and responses (adenylyl cyclase inhibition). They found evidence that the high-affinity state of the α_2-adrenergic receptor is an obligate intermediate in the process of G-protein activation (5) and estimated rate-constants for steps in the G_i activation cycle (37).

5.3 Spectroscopic methods

A very powerful use of direct agonist-binding to a G-protein coupled receptor has utilized a fluorescent peptide agonist for the neutrophil chemotaxis receptor (38, 39). With this ligand, G-protein-dependent agonist binding can be detected in real time. In permeabilized neutrophils, the existence of a pre-coupled RG complex was inferred by a kinetic analysis similar to that in Section 4.3.2 but with much more complete data (40).

Intrinsic protein fluorescence has also been used to measure activation of

G-proteins (41–43). All proteins have a certain amount of intrinsic fluorescence at approx. 280 nm, due to the presence of tryptophan. Conformational changes in the protein induced by GTP analogues, magnesium, and receptor activation result in marked changes in the intensity of this intrinsic fluorescence. An advantage of this method over extrinsic fluorphores is that no special fluorescent probes or chemical modification of the proteins is required. A disadvantage is that this approach is only applicable to purified proteins because virtually all proteins exhibit tryptophan fluorescence, and plasma membranes would have too much background noise due to other proteins. Studies of this sort on RG interactions have been limited to the rhodopsin–transducin system where large amounts of both 'receptor' and G-protein are available (43).

Extrinsic fluorescence labelling of rhodopsin and transducin (44) provided direct evidence via resonance energy transfer of an association between rhodopsin and transducin. Work has begun on extrinsic fluorescence probes of G-proteins (45) to allow this approach to be used with the smaller amounts of protein available for more traditional receptors and G-proteins.

6. Perturbing receptor/G-protein interactions: exogeneous agents

Exogeneous agents which perturb RG interactions such as guanine nucleotides, bacterial toxins, proteases, antibodies to G-proteins, and sulfhydryl reagents provide additional tools to understand RG mechanisms. These agents often modify one of the protein components in a more or less specific manner. By elucidating their mechanism of action the structural determinants of RG interactions can be more easily understood.

Pertussis and cholera toxins (which are described in detail in Chapter 2) are examples of relatively specific exogeneous agents. Pertussis toxin ADP-ribosylates and inactivates G_i, and cholera toxin ADP-ribosylates and activates G_s. Both toxins uncouple the G-protein from receptor. They can be used to determine which type of G-protein is interacting since treatment with these toxins abolishes high-affinity agonist-binding to receptors coupled to the appropriate G-protein. They can also be used to determine which type of G-protein is involved in a particular cellular mechanism. These toxins can only determine if a general type of G-protein is involved (i.e. G_i or G_o vs. G_s or other non-pertussis substrates such as G_p), and not the precise subtype. Both toxins are very useful for treating whole cells grown in tissue culture, but they can also be used to treat plasma membranes (46).

Two less specific reagents for interfering with RG interactions are proteases and sulfhydryl reagents. Proteases such as chymotrypsin uncouple receptors from G_i. When platelet plasma membranes are pre-treated with 10 µg/ml chymotrypsin both high-affinity α_2-adrenergic agonist-binding and adenylyl cyclase inhibition are reduced (47, 48). *N*-Ethylmaleimide (NEM) has been used to inactivate G_i-mediated responses in many systems (49, 50). In fact, NEM alkylates the same cysteine residue that is modified by pertussis toxin in the G_i-like protein G_o (51).

Antibodies to particular subtypes of the G-proteins are highly specific exogeneous agents. Details of the use of G-protein antibodies are described in Chapter 2. For example, antipeptide antibodies specific to the G_{i2} protein α subunit block δ opiate receptor-mediated GTPase activation (52) and adenylyl cyclase inhibition (53). Future studies with specific antibodies will be important in elucidating the specificity of receptor interactions with the multiplicity of G-protein types and subtypes.

References

1. Freissmuth, M., Casey, P. J., and Gilman, A. G. (1989). *FASEB J.*, **3**, 2125.
2. Birnbaumer, L. (1990). *FASEB J.*, **4**, 3178.
3. Gilman, A. G. (1987). *Annu. Rev. Biochem.*, **56**, 615.
4. Neubig, R. R. and Szamraj, O. (1986). *Biochim. Biophys. Acta*, **854**, 67.
5. Thomsen, W. J., Jacquez, J. A., and Neubig, R. R. (1988). *Mol. Pharmacol.*, **34**, 814.
6. Dalman, H. M., and Neubig, R. R. (1991). *J. Biol. Chem.*, **266**, 11025.
7. Ross, E. M., Maguire, M. E., Sturgill, T. W., Biltonen, R. L., and Gilman, A. G. (1977). *J. Biol. Chem.*, **252**, 5761.
8. Gerhardt, M. A., Wade, S. M., and Neubig, R. R. (1990). *Mol. Pharmacol.*, **38**, 214.
9. Pedersen, S. E., and Ross, E. M. (1982). *Proc. Natl. Acad. Sci. USA*, **79**, 7228.
10. Sternweis, P. C., Northup, J. K., Hanski, E., Schleifer, L. S., Smigel, M. D., and Gilman, A. G. (1981). *Adv. Cyclic. Nucleotide. Res.* **14**, 23.
11. Shorr, Robert G. L., Strohsacker, Mark W., Lavin, Thomas N., Lefkowitz, Robert J., and Caron, Marc G. (1982). *J. Biol. Chem.*, **257**, 12341.
12. Levitzki, Alexander (1985). *Biochim. Biophys. Acta*, **822**, 127.
13. Lin, M. C., Nicosia, S., Lad, P. M., and Rodbell, M. (1977). *J. Biol. Chem.*, **252**, 2790.
14. Wreggett, K. A. and De Lean, A. (1984). *Mol. Pharmacol.*, **26**, 214.
15. Costa, T. and Herz, A. (1989). *Proc. Natl. Acad. Sci. USA*, **86**, 7321.
16. Snavely, M. D. and Insel, P. A. (1982). *Mol. Pharmacol.*, **22**, 532.
17. Motulsky, H. J. and Ransnäs, L. A. (1987). *FASEB J.*, **1**, 365.
18. Munson, P. J. and Rodbard, D. (1980). *Anal. Biochem.*, **107**, 220.
19. Kim, M. H. and Neubig, R. R. (1985). *FEBS Lett.*, **192**, 321.
20. Citri, Y. and Schramm, M. (1980). *Nature*, **287**, 297.
21. Neubig, R. R., Gantzos, R. D., and Thomsen, W. J. (1988). *Biochemistry*, **27**, 2374.
22. Costa, T., Lang, J., Gless, C., and Herz, A. (1990). *Mol. Pharmacol.*, **37**, 383.
23. Abramson, S. N., McGonigle, P., and Molinoff, P. B. (1987). *Mol. Pharmacol.*, **31**, 103.

24. Limbird, L. E. (1986). *Cell surface receptors: A short course on theory and methods*. Martinus Nijhoff, Boston.
25. Kim, M. H., and Neubig, R. R. (1987). *Biochemistry*, **26**, 3664.
26. Taylor, P., and Insel, P. A. (1990). In *Principles of drug action: The basis of pharmacology* (ed. W. B. Pratt and P. Taylor), pp. 1–102, Churchill Livingstone, New York.
27. Gantzos, R. D., and Neubig, R. R. (1988). *Biochem. Pharmacol.*, **37**, 2815.
28. De Lean, A., Stadel, J. M., and Lefkowitz, R. J. (1980). *J. Biol. Chem.*, **255**, 7108.
29. Neubig, R. R., Gantzos, R. D., and Brasier, R. S. (1985). *Mol. Pharmacol.*, **28**, 475.
30. Ransnäs, L. A. and Insel, P. A. (1988). *J. Biol. Chem.*, **263**, 9482.
31. Boyd, N. D. and Cohen, J. B. (1980). *Biochemistry*, **19**, 5344.
32. Boyd, N. D. and Cohen, J. B. (1980). *Biochemistry*, **19**, 5353.
33. Foster, D. M. and Boston, R. C. (1982). In *Compartmental distribution of radiotracers* (ed. J. R. Robertson), pp. 73–142. CRC Press, Boca Raton, Florida.
34. Asano, T. and Ross, E. M. (1984). *Biochemistry*, **23**, 5467.
35. Brandt, D. R. and Ross, E. M. (1986). *J. Biol. Chem.*, **261**, 1656.
36. Tota, M. R. and Schimerlik, M. I. (1990). *Mol. Pharmacol.*, **37**, 996.
37. Thomsen, W. J. and Neubig, R. R. (1989). *Biochemistry*, **28**, 8778.
38. Sklar, L. A. (1987). *Annu. Rev. Biophys. Biophys. Chem.*, **16**, 479.
39. Sklar, L. A., Mueller, H., Omann, G., and Oades, Z. (1989). *J. Biol. Chem.*, **264**, 8483.
40. Fay, S. P., Posner, R. G., Swann, W. N., and Sklar, L. A. (1991). *Biochemistry*, **30**, 5066.
41. Higashijima, T., Ferguson, K. M., Sternweis, P. C., Ross, E. M., Smigel, M. D., and Gilman, A. G. (1987). *J. Biol. Chem.*, **262**, 752.
42. Phillips, W. J. and Cerione, R. A. (1988). *J. Biol. Chem.*, **263**, 15498.
43. Guy, P. M., Koland, J. G., and Cerione, R. A. (1990). *Biochemistry*, **29**, 6954.
44. Borochov-Neori, H. and Montal, M. (1989). *Biochemistry*, **28**, 1711.
45. Kwon, G. and Neubig, R. R. (1989). *Biophys. J.*, **55**, 58a. (Abstract)
46. Kurose, H., Katada, T., Amano, T., and Ui, M. (1983). *J. Biol. Chem.*, **258**, 4870.
47. Ferry, N., Adnot, S., Borsodi, A., Lacombe, M. L., Guellaen, G., and Hanoune, J. (1982). *Biochem. Biophys. Res. Commun.*, **108**, 708.
48. Periyasami, S. and Somani, P. (1987). *Biochem. Pharmacol.*, **36**, 3086.
49. Asano, T. and Ogasawara, N. (1986). *Mol. Pharmacol.*, **29**, 244.
50. Smith, M. M. and Harden, T. K. (1984). *J. Pharmacol. Exp. Ther.*, **228**, 425.
51. Hoshino, S., Kikkawa, S., Takahashi, K., Itoh, H., Kaziro, Y., Kawasaki, H., Suzuki, K., Katada, T., and Ui, M. (1990). *FEBS Lett.*, **276**, 227.
52. McKenzie, F. R., Kelly, E. C., Unson, C. G., Spiegel, A. M., and Milligan, G. (1988). *Biochem. J.*, **249**, 653.
53. McKenzie, F. R. and Milligan, G. (1990). *Biochem. J.*, **267**, 391.
54. Haga, K., Haga, T., and Ichiyama, A. (1986). *J. Biol. Chem.*, **261**, 10133.

Basic techniques to study
G-protein function

FERGUS R. McKENZIE

1. Introduction

After the initial discovery of the involvement of a G-protein in mediating the hormonal stimulation of adenylyl cyclase (1, 2), it is now apparent that an entire family of G-proteins exists which are required for the cellular signal transduction cascade (3–5). The classical G-proteins are heterotrimers consisting of non-identical α, β, and γ subunits. Although it is subject to debate, the α subunits seem to be responsible for both interaction with transmembrane receptors as well as with intracellular effectors (6).

All G-proteins which have been analysed in detail seem to function in the same manner (*Figure 1*), upon agonist occupation of receptor, the G-protein α subunit loses its bound GDP and binds the activating nucleotide, GTP. It is the dissociation of GDP from the α subunit which is the rate limiting step in G-protein activation (7). In the presence of Mg^{2+}, the G-protein with GTP bound may dissociate into the now active α subunit and the βγ subunits. Upon interaction with intracellular effector, the G-protein α subunit becomes de-activated by an intrinsic GTPase activity, cleaving the γ-position phosphate of GTP to restore GDP in the active site. The now-inactive G-protein can re-associate with both the βγ subunits and with receptor, allowing the cycle of activation to continue (3, 8). Examples of receptors which function through G-protein activation include those for both α- and β-adrenergic ligands, muscarinic receptors, and prostaglandin receptors in addition to a wide variety of peptide hormone receptors. The list of receptors known to function through G-protein activation presently exceeds 150 examples (5). An assortment of intracellular effectors have been identified as being G-protein-regulated, including adenylyl cyclase (9), cyclic GMP phosphodiesterases (8), phosphoinositidase C (10), and phospholipase A2 (11). G-proteins are also known to interact with a range of ion channels and are able to inhibit certain voltage-sensitive Ca^{2+} transients (12), as well as stimulating cardiac K^+ channels (13). In addition, less well characterized events such as secretion and exocytosis (14), olfactory transduction (15), and

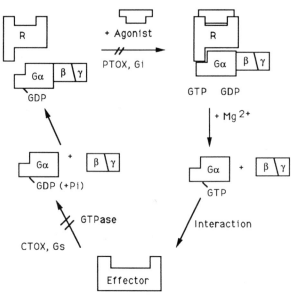

Figure 1. The G-protein cycle of receptor-stimulated activity. The explanation of this figure may be found in the text.

Na^+ channel activity (16) have been demonstrated to involve G-protein activation. The list of both receptors and effectors which require G-proteins to mediate their function is increasing all the time, and it is therefore likely that the list is not yet complete.

In view of the ubiquitous distribution of G-proteins in signal transduction systems, it has become of increasing importance that a laboratory can perform a variety of simple G-protein assays and thus easily gain a wealth of information. In addition to describing methods for the production of plasma membranes, upon which many G-protein assays are based, this chapter describes several techniques commonly used to study G-proteins which are easy to set up in the laboratory and do not require the purchase of specialized equipment.

2. Equipment required

2.1 Centrifuges

i. A high-speed centrifuge (ultracentrifuge) capable of producing up to 50 000 *g*. This should accept a fixed angle rotor (10-ml tube volume size).

ii. A bench centrifuge (microcentrifuge) capable of speeds up to 12 000 r.p.m. This should accept 1.5 ml Eppendorf-style tubes.

2.2 Homogenizers

i. A small homogenizer of the Potter-Elvehjem type having a glass mortar of approximately 10 ml capacity, and a Teflon pestle. The clearance should be about 0.1 to 0.2 mm.

ii. A polytron-style homogenizer with a cutting head of maximal diameter 15 mm.

2.3 Gel electrophoresis equipment

i. It is advisable to have a range of different gel electrophoresis plates since several of the techniques described rely on the use of longer than average gels for G-protein resolution. A suitable system is the Bio-Rad Protean II, which is easy to use and flexible in its applications.

2.4 Miscellaneous equipment

i. Gel drying system.

ii. Dark-room and autoradiography development facilities.

iii. A vacuum filtration apparatus suitable to filter up to 12 samples individually. Filters having a diameter of 2.5 cm.

iv. Scintillation counter.

v. Hand-held UV lamp, 254 nm, 150 W.

3. Plasma membrane production

Many current G-protein analysis techniques utilize cell-free systems, often relying on the production of cell plasma membranes. This is advantageous in that it simplifies G-protein experiments; however, it has a major drawback in that one has to decide if one wishes to perform assays in buffers which are similar in composition to either the intracellular or the extracellular millieu. In each case, the buffers will be quite different and this has to be taken into account when interpreting experiments in relation to physiological conditions. The methods detailed below are applicable to the production of plasma membranes from both cell-culture-derived material, as well as tissue obtained from whole animals. In both cases, the starting material should be stored at −70 °C prior to membrane production and thoroughly washed in PBS to remove any serum which may be present. Any remaining serum will act as a G-protein stimulator and interfere with the assays later described (17).

Protocol 1. Plasma membrane preparation from cells grown in artificial culture

All operations should be carried out at 4 °C.

1. Thaw the starting material in 5 vol. of ice-cold 10 mM Tris–HCl, 0.1 mM EDTA pH 7.5 (TE buffer), resuspend cells with the aid of a plastic Pasteur pipette.

2. Homogenize cells using a hand-held Potter-Elvehjem type homogenizer of approx. 15 ml capacity. 20 firm strokes are sufficient.[a]

3. Pass the homogenate through a fine-gauge syringe needle (26-G).

4. Centrifuge the homogenate at 500 g for 15 min to pellet unbroken cells and nuclei, leaving a membrane containing supernatant.

5. Collect plasma membranes by centrifugation of the supernatant at 40 000 g.

6. Wash the plasma membrane-containing pellet in 10 volumes of TE buffer and re-centrifuge at 40 000 g.

7. Resuspend plasma membranes in TE buffer to a final protein concentration of between 2 to 4 mg/ml, aliquot (approx. 200 µl fractions), store at −70 °C.

[a] Ensure that the plunger is a tight fit in the glass homogenizer, otherwise only a small proportion of cells will be disrupted.

The hypotonic buffer cell disruption methods detailed in *Protocols 1* and *2* produce a crude plasma membrane fraction which is sufficient for most membrane based G-protein assays. Unfortunately this is an inefficient method of plasma membrane production, since as much as 50% of the starting material is often not homogenized. The extent of homogenization may be visualized by comparing the pellets obtained after both the low- and high-speed spins. In the case of poor initial homogenization it is possible to re-homogenize and pool the high-speed pellets to obtain greater yield. There is little to be gained in further purification of the membrane fraction because the losses in both materials and preparation time outweigh a modest increase in membrane activity. On a day-to-day basis, purity and activity may be determined using a plasma membrane marker assay such as that for the ouabain sensitive (Na^+, K^+)ATPase. (18). Proteolysis of signal transduction components can often be a problem, particularly when using cells known to contain a variety of protein-degrading machinery such as the monocyte-derived HL60 and U937 cell lines. Since the protease activity in many tissues is poorly defined, a number of anti-protease cocktails have been cited in the literature. Commonly used protease inhibitors are listed in *Table 1*.

Table 1. Commonly used protease inhibitors

Compound	Concentration	Active against
Pepstatin	5 µg/ml	Lysosomal proteases
Leupeptin	10 µg/ml	Lysosomal proteases
Benzamidine hydrochloride	2 mM	Non-specific inhibitor
Bacitracin	0.1 mg/ml	Non-specific inhibitor
Phenylmethylsulfonyl fluoride	5 mM	Serine proteases[a]
EDTA or EGTA	10 mM	Ca^{2+} dependent proteases

[a] This must be made fresh daily.

The method described in *Protocol 1* is suitable for cell numbers of between 10^7 to 10^{12}, depending on the size of the cells; however, when producing cells from animal tissue, it is often advantageous to employ a slightly different protocol, as in *Protocol 2*.

Protocol 2. Plasma membrane production from whole animal tissue: suitable for between 1 to 25 g of tissue

1. Wash tissue in PBS to remove unwanted animal fluids, remove PBS and dry tissue lightly using blotting paper.
2. Cut up the tissue into small pieces with sharp scissors (about 2–3 mm square), or alternatively use a tissue chopper.
3. Add 5 volumes of TE buffer and homogenize using a Polytron-type homogenizer. Use no more than three 20-sec pulses at full speed. Follow steps 3 to 7 in *Protocol 1* above.

As the protease inhibitors listed in *Table 1* are usually expensive, their use is often confined to the initial homogenization steps, where they are added to the relevant buffers directly before use. It is worth checking individual membrane preparations to determine if indeed they are required. In addition to the use of antiproteases, it is often useful to perform the homogenization process rapidly since proteolytic activity often occurs at 0 to 4 °C.

4. The determination of GTPase enzymatic activity

G-proteins are enzymes which function to bind and hydrolyse GTP, producing GDP and inorganic phosphate. The rate of GTP hydrolysis is slow, the majority of G-proteins having turnover numbers (K_{cat}) in the region of 2–5 min^{-1} (3). However, the high concentration of G-proteins at the plasma membrane allows the measurement of GTPase activity under suitable conditions (19). In addition, it is often possible to measure receptor stimulation of the rate of GTP hydrolysis since receptors increase the rate of exchange of GDP on the α subunit for GTP and thereby increase the rate of GTP hydrolysis (20). Determination of G-protein-mediated GTPase activity is achieved by following the breakdown of $\gamma[^{32}P]GTP$ with the concomitant production of $[^{32}Pi]$. The $[^{32}Pi]$ is then separated from the $\gamma[^{32}P]GTP$, allowing the rate of GTP hydrolysis to be measured (7).

The original identification of G-proteins as GTP-hydrolysing enzymes was made in 1976 (19). The original assay may still be used, but several modifications are useful. The assay succeeds because it employs a low concentration of GTP (0.5 μM) along with the inhibition of non-specific nucleoside triphosphatases with adenosine $5'(\beta,\gamma,imino)$ triphosphate (App[NH]p). In addition, an ATP regenerating system is present (creatine phosphate, creatine phosphokinase) to suppress the transfer of ^{32}Pi liberated from $\gamma[^{32}P]GTP$ to ADP. Further reduction in the non-specific hydrolysis of GTP by ATPase enzymes is obtained by the inclusion of ouabain to inhibit the Na^+/K^+ ATPase. As with later procedures in this chapter. $MgCl_2$ is required in the assay both for the catalytic activity of the G-protein as well as G-protein dissociation (3, 20, 22). NaCl is required to attenuate the interaction of unoccupied receptor with the G-protein (21–23). Indeed, it is probably the interaction and activation of G-proteins by unoccupied receptors which accounts for the basal GTPase activity. In the complete absence of Na, the basal activity may be so high as to prevent the measurement of receptor stimulation of GTPase activity. Interestingly, the mechanism by which Na mediates this effect although at present unknown, may involve its binding to a regulatory site on the receptor (24).

Protocol 3. Agonist stimulation of GTPase activity

All manipulations to be carried out on ice.

1. Plasma membrane proteins (*Protocols 1* and *2*) are diluted in 10 mM Tris–HCl, 0.1 mM EDTA, pH 7.5 (TE buffer) to a final concentration of between 0.1 to 0.5 mg/ml, depending on origin and activity of sample.

2. An assay mixture is produced with the following reagents:

Reagent	*Concentration in mixture*	*Final concentration in assay*
App(NH)P	2 mM	1 mM
ATP	2 mM	1 mM
Ouabain	2 mM	1 mM
Creatine phosphate	20 mM	10 mM
Creatine phosphokinase	5 Units/ml	2.5 Units/ml

Sodium chloride	200 mM	100 mM
Magnesium chloride	10 mM	5 mM
Dithiothreitol	4 mM	2 mM
EDTA	0.2 mM	0.1 mM
Tris–HCl	20 mM	10 mM
$\gamma[^{32}P]GTP$	1 µM	0.5 µM[a]

Final pH of assay mixture is 7.5. An aliquot of this mixture is retained for radioactivity counting, such that the exact number of counts in each tube is known.

3. Aliquots of the above reaction mixture (50 µl) are added to Eppendorf tubes on ice, along with 20 µl of the diluted membrane protein and 10 µl of the appropriate drug. The final assay volume is 100 µl, which may be made up with water.[b]

4. Low-affinity hydrolysis of $\gamma[^{32}P]GTP$ (not G-protein mediated), is assessed by incubating parallel tubes in the presence of 100 µM GTP.

5. Blank values are determined by the replacement of membrane protein with TE buffer.

6. The reaction is initiated by transferral of the tubes to a 37 °C water bath.

7. After 20 min, the assay is terminated by removal of the tubes to ice, under these assay conditions, the hydrolysis of $\gamma[^{32}P]GTP$ on ice is negligible.

8. To both prevent further enzymatic activity and separate the free $[^{32}Pi]$ from the unhydrolysed $\gamma[^{32}P]GTP$, 900 µl of a 5% (w/v) activated charcoal slurry in 20 mM phosphoric acid (pH 2.3) is added to each tube, giving a total volume of 1000 µl.

9. Tubes are then centrifuged at 12 000 g for 20 min at 4 °C to pellet the charcoal along with unhydrolysed $\gamma[^{32}P]GTP$. The free $[^{32}Pi]$ is present in the supernatant.[c]

10. 500 µl aliquots of the supernatant fluid are removed from each tube and added to scintillation vials for radioactivity counting.

11. Radioactivity may be assessed by either liquid scintillation or Cerenkov counting.

[a] Since the radioactive GTP will be in trace amounts, it is necessary to add cold GTP up to the required concentration.
[b] Drugs may be made up either in water or in appropriate solvent.
[c] This is best carried out in a 4 °C cold room since heating tubes to room temperature encourages charcoal to stick to the side of the Eppendorf tube and makes it more difficult to remove the supernatant cleanly.

4.1 Calculation of GTPase assay results

Each assay tube contained 50 pmol of GTP (0.5 μM) in addition to approximately 50 000 c.p.m. This should be calculated each time by counting 50 μl aliquots of the assay mixture. Hence the specific activity of the GTP is approx. 1000 c.p.m. per picomole. After subtracting the blank values (no membranes), the rate of hydrolysis of GTP is calculated by

$$\frac{C}{S.A.} \times 2 \times \frac{1000}{P} \times \frac{1}{T}$$

where:
C = the counts in the 500 μl sample
$S.A.$ = the specific activity of the GTP
P = the amount of protein present in micrograms
T = the time of assay

This will give the rate of hydrolysis of GTP in pmol/min/mg of membrane protein.

When employing membranes produced from tissue which is greatly enriched in G-protein, such as brain, 2 μg of protein per tube is sufficient. However, for less active plasma membranes such as those derived from smooth muscle, 10 μg will be required. The above-described assay should be linear up to 25 min; however, this should be measured beforehand. In addition, no more than 15% of the GTP should be hydrolysed by any sample tube or else substrate limitation will affect the results obtained and produced non-sigmoidal V against S plots.

The GTPase assay has been widely used in a variety of tissues (see *Figure 2* for examples). However, it should be noted that this technique fails to demonstrate receptor stimulation of G_s in all model systems studied with the exception of the avian erythrocyte and the platelet. The reason for this is unclear but may be related to the significantly lower rate of GTPase activity of G_s, compared to the pertussis-toxin-sensitive G-proteins. In the platelet and the avian erythrocyte, however, it is known that G_s is present in much higher amounts in these two cell types and that both cell types express a large number of β-adrenergic receptors which are capable of G_s stimulation. In addition to limitations of the GTPase assay in the study of G_s, this technique is often of little use in the analysis of receptors which function through the activation of pertussis-toxin-insensitive G-proteins. This may be a reflection that the pertussis-toxin-insensitive G-proteins are expressed in low amounts, or that they hydrolyse GTP at a much slower rate than the pertussis-toxin-sensitive G-proteins. Interestingly, studies on a pertussis-toxin-insensitive G-protein, G_z, expressed in and purified from *E. coli* demonstrate that the rate

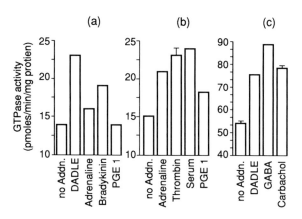

Figure 2. Basal and stimulated GTPase activity in a variety of cell membranes. The basal and receptor-stimulated GTPase activity was determined as described in *Protocol 3*, in a variety of plasma membranes. **A**: NG108-15 cell membranes (5 µg). **B**: Platelet membranes (8 µg), and **C**: Rat brain membranes (2 µg). Ligands were used at the following concentrations; DADLE (D-Ala$_2$ leucine enkephalin), adrenaline, bradykinin, PGE1, GABA (γ-amino butyric acid) all at 10 µM. Carbachol at 0.1 mM, thrombin at 5 U/ml and serum at 10% (v/v) final. The basal GTPase activity is represented by no addition.

of GTP hydrolysis by this G-protein is indeed slow, perhaps 200-fold slower than that of G$_s$α and G$_o$α (25, 26). It may be speculated that other members of the family of toxin-insensitive G-proteins share the same characteristic.

As a final point, the stimulation of GTPase activity produced by a given receptor may be a function of the number of receptors which are expressed by a given cell, such that the greater the number of receptors present, the greater the G-protein stimulation obtained. One would expect that at very high receptor numbers, saturation of GTPase stimulation is produced, however this has yet to be rigorously examined. This is in direct contrast to the results obtained in functional assays where individual receptor species are equally able to fully stimulate or inhibit an intracellular effector, but give very different stimulations of GTPase activity (F. R. McKenzie and G. Milligan, unpublished observations).

5. Gel electrophoresis of G-proteins

A continuing problem in G-protein analysis is that of G-protein subtype identification. In any given tissue, a complex variety of highly homologous G-proteins may be expressed. If we first examine the pertussis-toxin-sensitive G-proteins, then at present, of the 16 cDNA sequences corresponding to distinct α-subunits which have been isolated, eight sequences have the prerequisite recognition motif for pertussis toxin catalysed ADP-ribosylation, and are likely to be substrates for this toxin when translated into protein products *in vivo*. All of these substrates will have molecular masses of between 39 to 41 kDa under standard conditions of SDS-polyacrylamide gel electrophoresis. Hence a broad band of [^{32}P]ADP-ribose incorporation (*Figure 5*, lane *a*) may be composed of several pertussis toxin sensitive G-proteins.

To obtain better resolution of the pertussis toxin sensitive G-proteins by the use of the discontinuous SDS-PAGE system developed by Laemmli (26), several methods are applicable, depending on the addition of urea or not, as detailed in *Table 2*. A more detailed explanation of the techniques available for the gel electrophoresis of proteins is available (27).

Table 2. Resolving gel compositions for G-protein resolution

Resolving gel	Gel dimensions height × width × thickness (mm)
Gels without urea:	
12.5% acrylamide, 0.06% bisacrylamide	160 × 140 × 1.5
Gels with urea:	
(a) 4 M urea, 9% acrylamide, 0.26 bisacrylamide	120 × 140 × 1.5
(b) 6 M urea, 9% acrylamide, 0.26 bisacrylamide	120 × 140 × 1.5

Notes
(a) The above dimensions are for the resolving gel, not for the glass plates to be employed.
(b) In all cases, it is not necessary to employ special stacking gels; however, a stacking gel of less than 1 cm is not advised.
(c) Urea should be de-ionized prior to use using a mixed-bed de-ionizer such as AG 501-X8, obtainable from Bio-Rad Ltd.

Examples of the resolution possible by the above described SDS-PAGE resolving gel systems are shown in *Figure 3*. Samples (*a*) and (*c*) are identical; however, the presence of the three G-protein subtypes is masked in lane (*a*). It should be noted that even under resolving conditions (12.5% acrylamide, 0.06% bisacylamide, 160 mm gel height), it is often not possible to adequately resolve G_i3. To circumvent this problem, urea gels may be required. On both 10% and 12.5% acrylamide gels, $G_o\alpha$ migrates as a single species (*Figure 3*, lane *c*). However, it is possible to resolve different forms of $G_o\alpha$ by including urea in the resolving gel. In the presence of 4 M urea, $G_o\alpha$ may be resolved into two distinct components (*Figure 4*, lane *a*). Interestingly, in the presence of 6 M urea, three forms of $G_o\alpha$ may be identified (*Figure 4*, lane *b*). Although at least two forms of G_o are known to exist (28), it is not possible to determine whether the third band visualized on 6 M urea gels is indeed

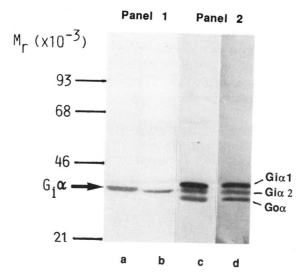

Figure 3. G-protein resolution using polyacrylamide gels. Plasma membranes from either NG108-15 cells (50 μg), (lane **b**), or from rat brain (40 μg) (lanes **a, c, d**) were resolved on either 10% acrylamide, 0.25% bisacrylamide gels (*Panel 1*), or on 12.5% acrylamide, 0.06% bisacrylamide gels (*Panel 2*) (See *Table 2*). The samples in lane (**d**) were alkylated with NEM (*Protocol 4*). Gels were then Western blotted with a mixture of antisera known to recognize G_i1, G_i2, and G_o. (Kindly provided by Dr G. Milligan, University of Glasgow.) Identification was made by comparison against a range of known standards (not shown).

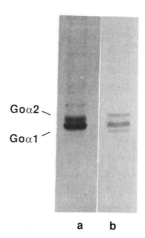

Figure 4. G-protein resolution using urea gels. Plasma membranes (50 μg) from rat brain (*Protocol 2*) were resolved on polyacrylamide gels (9% acrylamide, 0.26 bisacrylamide) in the presence of either 4 M (lane **a**) or 6 M (lane **b**) urea. Gels were then Western blotted with an antiserum which specifically identifies $G_o\alpha$. (Kindly provided by Dr G. Milligan, University of Glasgow.)

another form of G_o, or if it is a modified form of a previously recognized G_o. Caution should be maintained when attempting to identify the various G-protein species expressed by a given tissue when using urea gels, since gels with a high urea content do not allow protein migration strictly according to molecular mass. For example, the G-protein β subunit (35 kDa mass) migrates more slowly in urea containing gels than the G-protein α subunits (39–45 kDa mass).

If the above-detailed resolving gel systems do not provide sufficient separation of the G-protein α subunits, then greater resolution may be obtained by the alkylation of G-protein α subunits prior to gel electro-phoresis. This technique is based on the alkylation of various cysteine residues in the α subunit, by N-ethylmaleimide (NEM), as detailed in *Protocol 4*.

Protocol 4. Alkylation of G-protein α subunits by NEM

1. Dilute membrane samples to the required protein concentration and centrifuge at 12 000 g for 10 min at 4 °C. Discard supernatant.

2. Resuspend pellet in 20 µl of 10 mM Tris–HCl, 0.1 mM EDTA pH 7.5.

3. Add 10 µl of 5% (w/v) SDS, 50 mM DTT to each sample, mix and place in a boiling-water bath for 15 min.

4. Cool samples and add 10 µl of freshly made 100 mM NEM.[a]

5. After 15 min at room temperature, add 20 µl of Laemmli buffer (5 M urea, 0.17 M SDS, 0.4 M DTT, 50 mM Tris–HCl pH 8.0) to each sample, leaving the sample ready for gel electrophoresis.

[a] Final NEM concentration in assay is 25 mM.

Although the α subunits of $G_i1,2,3$, and G_o contain 10 cysteine residues, only three are apparently susceptible to alkylation by NEM. However, in our laboratory it has been noted that NEM treatment for up to 45 min will produce a heterogeneous pool of an individual G-protein species, leading to an indistinct broad band of the α subunit upon gel electrophoresis; hence attention must be paid to the time of alkylation. The effect of NEM alkylation is to make the G-protein α subunits migrate more slowly in SDS-PAGE. However, G_o is least affected by this treatment, with the result that it is possible to obtain greater resolution of $G_o\alpha$ from the α subunits of $G_i1,2$, and 3. In contrast, after NEM treatment, $G_i1,2$, and 3 do not alter their mobility relative to each other (see *Figure 3*, lane *d*).

6. Two-dimensional gel electrophoresis of G-proteins

It is possible to use two-dimensional electrophoresis systems to resolve G-proteins and this has indeed been employed by several groups (29–31); however, in practice this is often quite difficult to achieve. The main reason for difficulty seems to lie in the poor ability of G-protein α subunits to enter the first dimensional gel; perhaps 40% of the G-protein does not enter (I. Mullaney, F. R. McKenzie, and G. Milligan, unpublished observations). The reason for this is unclear; however, it may be possible to circumvent the problem by casting the first dimensional gel with the G-protein sample present in the gel mix. The G-protein will then be evenly spread throughout the length of the first dimensional gel and should focus normally.

7. G-protein ADP-ribosylating toxins

An enormous asset to G-protein study was the discovery of two bacterial exotoxins which are able to catalyse the covalent modification of G_s and G_i. G_i refers to the 'G_i-like' G-proteins, including $G_i1,2,3$, $G_o1,2$, and Td1 and 2. Cholera toxin, isolated from cultures of *Vibrio cholerae*, is able to catalyse the mono ADP-ribosylation of $G_s\alpha$, using NAD as substrate (32). The target site is an arginine residue, Arg 187/188, or 201/202; the exact number of this residue depends on which of the four splice variants of $G_s\alpha$ is under study (33). The functional effect of this modification is to reduce the ability of $G_s\alpha$ to hydrolyse GTP. $G_s\alpha$ is thus maintained in its active conformation (34). In cell systems, cholera toxin treatment *in vivo* will lead to an increase in adenylyl cyclase activity and therefore an elevation of intracellular cyclic AMP levels.

In a similar manner, pertussis toxin, isolated from cultures of *Bordetella pertussis*, is able to catalyse the mono-ADP-ribosylation of members of the G_i family of G-proteins (35, 36). The target site of modification of 'G_i-like' G-proteins by pertussis toxin is a cysteine residue four amino acids removed from the C-terminus. In contrast to ADP-ribosylation of G_s, when G_i is ADP-ribosylated by pertussis toxin, receptor activation of G_i is attenuated (37, 38). This will maintain G_i in its inactive $\alpha\beta\gamma$ heterotrimeric conformation. The consequence of pertussis toxin catalysed ADP-ribosylation *in vivo* will again be an increase in cAMP within the cell, although indirectly, since adenylyl cyclase is no longer under inhibitory control (38).

In many cell lines, it is not possible to measure a difference in intracellular cAMP after pertussis toxin treatment, even though G_i has been fully ADP-ribosylated. This may be due to a low level of expression of G_s such that removal of the inhibitory action of G_i has no functional consequence to the cell. The protocol for ADP-ribosylation *in vivo* using both pertussis and cholera toxins is detailed in *Protocol 5*. This is applicable to cells grown in artificial culture conditions.

43

Protocol 5. Toxin treatment of cells *in vivo*.

1. When cells are nearly confluent (approx. 90%), add the required concentration of either pertussis or cholera toxin, or both, to fresh cell culture media.

2. Sterile filter the media and add to cells, add similar media containing toxin vehicle to a set of control cells.

3. After the required incubation time, dispose of cell media and harvest cells as normal, prior to membrane production.

4. For intracellular cyclic AMP measurement, simply remove toxin containing media and add fresh media.[a]

---

[a] Often intracellular cyclic AMP levels will be measured as a function of time, in which case it is not necessary to change the growth media prior to the cAMP assay.

With regard to the concentrations of toxins required, it is usually advisable to treat cells with a high concentration of toxin for a shorter period of time, although this is obviously more expensive in terms of toxin usage. A suitable starting point for pertussis and cholera toxin is noted below.

Toxin		*Concentration* (ng/ml)	*Time of incubation* (hours)
Pertussis toxin	A	10	18
	B	100	4
Cholera toxin	A	50	18
	B	150	4

For cells which will be used for membrane production, a longer time of treatment is preferable. For cells which shall be used for functional studies *in vivo*, a shorter time-course is preferable. In both cases it is necessary to determine whether the time of treatment has been sufficient to fully ADP-ribosylate the G-proteins under study. In the case of pertussis-toxin-sensitive G-proteins, this is usually achieved by rechallenging membranes produced from cells treated *in vivo* using fresh toxin and a radioactive substrate, [^{32}P]NAD, as detailed in *Protocol 6*. Usually in excess of 90% of the available pertussis toxin substrates can be ADP-ribosylated by pertussis toxin treatment *in vivo*. In the case of cholera toxin catalysed ADP-ribosylation, it is not possible to ascertain if the entire pool of G_s has been ADP-ribosylated, since cholera toxin treatment *in vivo* has been demonstrated to promote the loss of $G_s\alpha$ from the plasma membrane (39, 40).

There are notable exceptions to use of both cholera and pertussis toxin treatment *in vivo*. In particular, it is not possible to pre-treat whole platelets with pertussis toxin. The toxin probably fails to enter the platelet due to the lack of expression of a specific receptor required for toxin binding. To circumvent this problem, functional studies on inhibitory G-proteins in platelets are usually performed on platelet membranes (*Protocol 1*), the membranes being pre-treated with a high concentration of pertussis toxin (up to 250 μg of platelet membranes are incubated with 40 μg/ml of DTT activated pertussis toxin (add an equal volume of 100 mM DTT to the toxin, leave at room temperature for 60 min) for 30 min at 30 °C, in the presence of 15 mM Hepes pH 8.0, 10 mM MgCl$_2$, 2 mM EDTA, 2 mM DTT, 20 mM thymidine and 10 μM NAD. After this time, a high proportion of the 'G$_i$-like' G-proteins will be ADP-ribosylated. The membranes can then be washed, centrifuged and resuspended for functional assays. Alternatively, if whole platelet studies are required, it is possible to permeabilize platelets with saponin, thus allowing access of the pertussis toxin, as detailed below.

Protocol 6. ADP-ribosylation of saponin permeabilized platelets

1. Whole platelets, at a concentration of approximately 1.5 × 10^9 cells/ml in TE buffer are combined with 3 vol. of the following reagents:

- 15 μg/ml DTT-activated pertussis toxin
- 160 mM KCl
- 27 mM Hepes pH 7.1
- 5 mM MgCl$_2$
- 17 μM EGTA
- 67 μM NAD

2. Permeabilization is initiated by the addition of 3 vol. of saponin at a concentration of 20 μg/ml. Cells are incubated for 60 min at 30 °C.

3. Following incubation, the permeabilized platelets are collected by brief centrifugation (30 sec) in a microcentrifuge, platelets are then re-suspended in TE buffer and are ready for either functional assay or membrane production.

As with many other signal transduction components, it is possible to study G-protein function in the *Xenopus* oocyte. However, difficulty is often encountered when attempting to treat oocytes with ADP-ribosylating toxins; as with platelets, this may be due to difficulty in the toxin entering the cell. This problem may be circumvented by using a high concentration of toxin.

7.1 *In vitro* ADP-ribosylation of G-proteins

In the presence of a radiolabelled substrate, [^{32}P]NAD, and appropriate assay conditions (see *Protocol 7*), both cholera and pertussis toxins will catalyse the incorporation of [^{32}P]ADP-ribose into their respective G-protein substrates (see *Figure 5*). The G-proteins thus radiolabelled may then be separated by the appropriate SDS-PAGE G-protein-resolving gel. This allows the identification of the G-protein subtypes expressed in a given tissue. In addition it should be possible to immunoprecipitate the various radiolabelled G-protein subtypes with G-protein specific antisera, aiding identification.

Protocol 7. *In vitro* ADP-ribosylation with pertussis and cholera toxins

1. Membranes to be ADP-ribosylated (see *Protocol 1*) are diluted in 10 mM Tris–HCl, 0.1 mM EDTA, pH 7.5, to a protein concentration of between 1 to 3 mg/ml.[a]

2. 20 μl aliquots of membranes are assayed in a final volume of 50 μl containing the following:
   - 3 μM [^{32}P]NAD (4 × 10^6 c.p.m.)
   - 250 mM potassium phosphate buffer, pH 7.0[b]
   - 20 mM thymidine
   - 1 mM ATP pH 7.0
   - 20 mM arginine hydrochloride
   - 100 μM GTP pH 7.0

3. The appropriate toxin is then added at a final concentration of 10 μg/ml (pertussis toxin), or 50 μg/ml (cholera toxin). Both toxins are pre-activated by treatment with 50 mM DTT for 60 min at room temperature prior to assay.[c]

4. The assay is initiated by transfer to a 37 °C water bath for 60 min. After this time the assay is terminated by removal to ice.

[a] 20 μg of protein (Lowry assay) is sufficient for most cell membranes
[b] ADP-ribosylating toxins have a requirement for high ionic strength.
[c] *In vivo* assays require the holotoxin conformation of both cholera and pertussis toxins, *in vitro* assays require the active conformation, hence the use of DTT.

Often after *in vitro* ADP-ribosylation, the samples are to be resolved under SDS-PAGE as described in Section 5. In this case, after ADP-ribosylation as described in *Protocol 7*, samples are prepared for gel electrophoresis by sodium deoxycholate/trichloroacetic acid precipitation, as detailed in *Protocol 8*.

Protocol 8. Preparing ADP-ribosylated samples for PAGE

1. To ADP-ribosylated samples (50 μl total volume), add 6.25 μl of 2% (w/v) sodium deoxycholate.

2. Add 750 µl of water, followed by 250 µl of 24% (w/v) trichloroacetic acid.

3. Samples are centrifuged at 12 000 g for 20 min at 4 °C. This will pellet the proteins.

4. Supernatants are removed and the pellets resuspended in 20 µl of 1 M Tris base, followed by 20 µl of Laemmli buffer (see *Protocol 4*). The samples are now ready for SDS-PAGE.[a]

---

[a] Samples should not be heated as this may promote G-protein aggregation and reduce resolution of the G-protein α subunits.

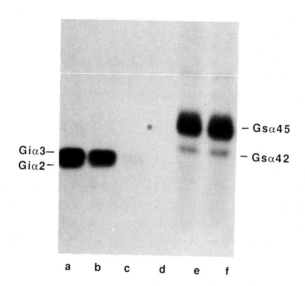

Figure 5. Pertussis and cholera toxin-catalysed [^{32}P]ADP-ribosylation of G_s and G_i. Membranes (30 µg) from either control CCL39 fibroblasts (Chinese hamster lung fibroblast derived) (lanes **a, b, e, f**), or from CCL39 cells pre-treated *in vivo* with either pertussis toxin (lane **c**) or cholera toxin (lane **d**) (*Protocol 5*), were ADP-ribosylated *in vitro* (*Protocol 7*). Samples were resolved on a 10% acrylamide, 0.26 bisacrylamide gel, and the gel obtained dried down under vacuum and exposed to X-ray film for 24 h.

In membranes produced from hamster CCL39 fibroblast cells (*Protocol 1*), the pertussis-toxin-catalysed [^{32}P]ADP-ribosylation of 'G_i' produces a broad band (*Figure 5*, lanes *a,b*), corresponding to a recombination of G_i3 and G_i2. By comparison, cholera toxin-catalysed ADP-ribosylation results in the incorporation of radioactivity into two forms of G_s (*Figure 5*, lanes *e, f*). The heterotrimeric αβγ conformation of G_i represents the best substrate for pertussis toxin catalysed modification; the free α subunit being a poor substrate for ADP-ribosylation. In contrast, cholera toxin prefers the free α subunit of G_s, the heterotrimeric αβγ conformation being a poor substrate. In addition, there are notable differences in the affinity of purified α subunits for the βγ subunits, a facet which has allowed purification and separation of highly homologous α subunits (43). This suggests that the various α subunits might not all be in the heterotrimeric state under the same conditions, depending on the affinity of their association with βγ subunits. This will lead to differences in the ADP-ribosylation pattern produced by both pertussis and cholera toxins when acting on both G_i and G_s respectively (44, 45). In the case of G_s, the ability of cholera toxin to ADP-ribosylate this G-protein is dependent on the presence of an additional co-factor, which has been termed ARF, (ADP-ribosylating factor) (46). For this reason, cholera toxin is unable to ADP-ribosylate purified G_s (unless ARF is added). It now appears that ARF may be one of a variety of ADP-ribosylating factors and the extent of ADP-ribosylation obtained from a variety of tissues will be dependent on the presence and amount of ARF present in a given tissue. Bearing these caveats in mind, it could be argued that it is permissible to quantify G-proteins using ADP-ribosylation within a given tissue; however, between tissues, this assay is subject to too many variables to be of use in quantitation.

8. G-protein identification using a photoreactive GTP analogue, [α³²P]GTP azidoanilide

The interaction of receptors with G-proteins is often demonstrated functionally by examining the effects of receptor agonists on high-affinity GTPase activity (see *Protocol 3*). Functional assays of this nature are based on the ability of a receptor occupied by agonist to stimulate the off rate of GDP from the G-protein α subunit. In the presence of GTP or a suitable analogue of GTP, the G-protein nucleotide binding site will be filled with the activating nucleotide.

It is possible to take advantage of the agonist–receptor driven selective displacement of GDP for GTP on the α subunit by the use of certain photoreactive nucleotides. Indeed, such nucleotides were originally used in the identification of G_s, the stimulatory G-protein (2, 45, 47). Subsequently, GTP-azidoanilide (*Figure 6*) has been used to identify and characterize various G-proteins including G_s, 'G_i-like' G-proteins and transducin (48, 49). The advantage of this method of G-protein detection is that it should be possible to selectively stimulate the release of bound GDP, followed by the binding of GTP-azidoanilide, only in the species of G-protein(s) which interact with the agonist-occupied receptor under study. After brief exposure to UV light of particular wavelength, the GTP-azidoanilide will covalently cross-link to the G-protein to which it has bound. Hence, by utilizing GTP-azidoanilide which is labelled with ³²P in either the α, β, or γ position, it should be possible to obtain agonist stimulation of incorporation of radioactivity into a particular class of G-protein(s). Following adequate resolution of these G-proteins under SDS-PAGE (see Section 4), it should then be possible to identify the G-proteins involved. The latter is absolutely crucial to this technique. Although GTP-azidoanilide is commercially available (8-azido-GTP), it is often better to produce the compound 'in house'. The protocols available for production of azido-GTP have been summarized in an excellent review (50).

In assays designed to examine receptor–G-protein specificity, a common problem of using non-hydrolysable analogues of GTP to activate G-proteins is that the rate of binding of these analogues is too rapid to allow agonist-induced increases in the binding rate to be measured (see *Protocol 10*). As with the activating nucleotides GTPγS and Gpp(NH)p, GTP-azidoanilide is resistant to hydrolysis and produces a persistent activation of G-proteins. However, by contrast, in the absence of agonist the rate of binding of GTP-azidoanilide to the G-protein α subunit seems to be considerably slower than that of both GTPγS and Gpp(NH)p. It is this facet which may allow the identification of G-proteins which can interact with specific receptors using [α-^{32}P]azidoanilide-GTP. Several groups have reported such an observation in a variety of different signal transduction model systems (51, 52).

Protocol 9. Photoaffinity labelling of G-proteins

All manipulations are carried out on ice.

1. Plasma membranes (*Protocol 1*) are diluted in 10 mM Tris, 0.1 mM EDTA pH 7.5, to a concentration of 2.5 mg/ml. The following assay buffer is then produced;

Solution	*Concentration in buffer*	*Final concentration in assay*
EDTA	0.3 mM	0.1 mM
MgCl$_2$	15 mM	5 mM
Benzamidine	3 mM	1 mM
NaCl	30 mM	10 mM
GDP	9 μM	3 μM
Hepes	90 mM	30 mM

   The final pH of this buffer should be 7.4.

2. To a darkened Eppendorf tube, add 20 μl of the above assay buffer to 20 μl of the diluted plasma membranes with or without 6 μl of receptor agonist(s). Tube are then incubated for 3 min at 30 °C. After this pre-incubation, the tubes are removed to ice.

3. Add 10 μl of [α-^{32}P] azidoanilide-GTP (1 μCi/tube), at a final concentration of between 3 to 50 nM. Final assay volume is 60 μl. Incubate tubes for a further 3 min at 30 °C. Stop assay by removal to ice.

Protocol 9. *Continued*

4. Remove unbound [α-^{32}P] azidoanilide-GTP by centrifugation of the tubes at 12 000 *g* for 5 min at 4 °C. Discard supernatant and resuspend pelleted membranes in a modified version of the above assay buffer (with the addition of 2 mM DTT and without GDP).

5. Membrane suspensions are then irradiated at 4 °C for 10 sec at a distance of 3 cm with a UV lamp (254 nm, 150 W). Membranes are again centrifuged and pellets dissolved in 20 μl of Laemmli sample buffer (see *Protocol 4*), ready for gel electrophoresis (Section 4).

a The [α-^{32}P] azidoanilide-GTP degrades upon contact with light and must therefore be kept shielded from light wherever possible.

The ability of the technique set out in *Protocol 9* to label G-proteins in a satisfactory manner (*Figure 7*), depends on a variety of factors known to affect G-protein function either directly or indirectly. First of all, the normal concentration of GTP in a whole cell is of the order of 1 mM, with GDP being present in at least tenfold lower concentrations. Since the K_M for GTP of most G-protein α subunits is in the range of 0.1 to 0.5 μM (3), this means that under normal physiological conditions the G-protein could easily be saturated with guanine nucleotides and active at all times. However, this is known not to occur, and it is the function of the receptor to promote the G-protein conformational change which determines whether the G-protein becomes activated or remains in an inactive state. The receptor achieves this by promoting the rate of release of GDP from the α subunit (7). Hence in assays designed to study receptor activation of G-proteins, it is necessary for GDP to be present such that the receptor can promote the release of this nucleotide and thus allow G-protein activation. In the above instance, this may be visualized as an increase in the rate of binding of [^{32}P] azidoanilide-GTP. Second, as discussed earlier, it is important to include Mg^{2+} (in the form of the chloride salt) in the assay. The function of this ion is unclear, however it appears to function by promoting the ability of the receptor to interact with

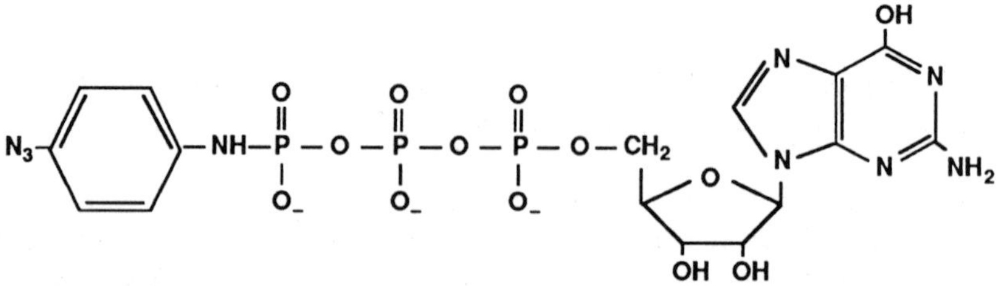

Figure 6. The structure of azidoanilide-GTP.

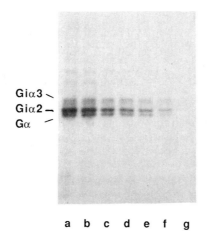

Giα3

Giα2

Gα

a b c d e f g

Figure 7. [^{32}P]azidoanilide-GTP labelling of G-proteins. Plasma membranes (50 µg) from CCL39 cells were labelled with [^{32}P] azidoanilide-GTP exactly as described in *Protocol 9*, in response to increasing amounts of GDP. Lane (**a**), no GDP, lane **b**, 0.1 µM GDP, lane **c**, 1 µM GDP, lane **d**, 3 µM GDP, lane **e**, 10 µM GDP, lane **f**, 30 µM GDP, lane **g**, no membranes.

the G-protein (22). Hence the inclusion of a high concentration of MgCl$_2$ in the assay (approximately 10 mM) will increase the probability of obtaining receptor-mediated G-protein stimulation. Lastly, it is appreciated that an empty receptor can interact with and activate G-proteins (20, 21). This will produce a high basal activity in any assay designed to look at the effects of receptor stimulation of G-proteins. It is possible to attenuate the 'empty-receptor effect' by the inclusion of a high concentration of NaCl (30 mM). The normal intracellular Na$^+$ concentration is in the region of 15 mM. The mechanism by which Na$^+$ mediates this effect seems to be directly on the receptor (24). It is therefore advised that when setting up assays to monitor the binding of [^{32}P] azidoanilide-GTP, initial experiments should involve concentration-dependent analysis of the effects of each of GDP, MgCl$_2$, and NaCl.

9. Receptor-regulated binding of guanosine-5'-O-(3-thiotriphosphate), ([^{35}S]GTPγS)

Initial studies on the mechanism of action of G-proteins involved the reconstitution of purified G-proteins, along with other protein components of the signal transduction cascade, into artificial phospholipid vesicles (53–55). By utilizing both purified G$_s$ and β-adrenergic receptor, it was possible to demonstrate a functional interaction between receptor and G-protein (see Chapter 1). This was characterized by both receptor-mediated stimulation of GTPase activity, as well as the stimulation of binding of the poorly-hydrolysable GTP analogue, GTPγS (*Figure 8*) (20). It is possible to apply both techniques to plasma membrane systems, although the latter is less frequently employed. However, by examining the effect of a receptor–agonist complex on the rate of binding of [35S]GTPγS (57, 58) it is possible to determine whether a given receptor class functions through the activation of a G-protein species which is either sensitive or insensitive to the actions of both cholera and pertussis toxins.

Protocol 10. Binding of [35S]GTPγS

All manipulations to be carried out on ice.

1. Membranes are diluted in 10 mM Tris–HCl, 0.1 mM EDTA pH 7.5, to between 1.25 and 2.5 mg/ml, to give a final concentration in the assay of between 25 to 50 μg/ml.

2. Make an assay mixture containing the following:

Reagent	Concentration in mixture	Final concentration in assay
Tris–HCl	20 mM	10 mM
MgCl$_2$	10 mM	5 mM
EDTA	0.2 mM	0.1 mM
DTT	2 mM	1 mM
NaCl	300 mM	150 mM
GDP	2 μM	1 μM

   the final pH is 7.5. The final assay volume is 100 μl.

3. To Eppendorf tubes on ice, add 50 μl of the above assay mixture to each tube followed by 10 μl of [35S]GTPγS (0.5 to 1 nM, approx. 50 000 c.p.m.).

4. Non-specific values are determined by the addition of 10 μl of 100 μM GTPγS^a (10 μM final). Blank values are determined by replacing membrane protein with TE buffer.

5. Agonist stimulation is achieved by the addition of 10 μl of appropriate agonist.

6. The assay is started by the addition of 20 μl of the diluted membrane protein, followed by rapid vortexing and placement in a 25 °C water bath.

7. The reaction is terminated at various time points by rapid filtration through Whatman GF/C glass-fibre filters under vacuum. Filters are then washed three times with 5 ml of 10 mM Tris–HCl, 0.1 mM EDTA pH 7.5 (15 ml total).

8. For radioactivity counting, filters are placed in a scintillation vial along with 20 ml of scintillation fluid. After being shaken well and left in a dark cupboard for at least 12 h, samples are counted.

a Non-specific binding is usually extremely low, often only 1% of the total [^{35}S]GTPγS.

For several membrane types, it is extremely difficult to follow the rate of binding of [^{35}S]GTPγS binding using the technique described in *Protocol 10*, since the binding is too rapid. Several techniques are available to reduce the rate of binding of this nucleotide including changing the incubation temperature. At 4 °C, the assay will be considerably slower. In addition, it is possible to reduce the basal rate of binding (no agonist), by the addition of NaCl (see earlier). As with the binding of [^{32}P] azidoanilide-GTP, it is often not possible to measure agonist-stimulation of the rate of binding of [^{35}S]GTPγS in the absence of GDP, the greatest stimulation being obtained in the presence of millimolar GDP (58). It should be noted that the effect of an agonist in the above assay is not to increase the maximal binding of [^{35}S]GTPγS, but is only to alter the rate at which binding equilibrium is achieved. This method is less useful for G-protein identification, since [^{35}S]GTPγS will bind to the entire complement of G-proteins in a given membrane preparation. If the receptor of interest functions through the activation of one species of G-protein and that G-protein is only a minor component of the total number of G-proteins present, then it will be extremely difficult to measure an increase in the rate of binding of [^{35}S]GTPγS over the background 'noise'.

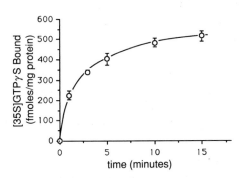

Figure 8. [^{35}S]GTPγS binding to G-proteins. The binding of [^{35}S]GTPγS to plasma membranes (50 μg) from NG108-15 cells was determined exactly as described in *Protocol 10*, at various time points (0, 1, 3, 5, 10, and 15 min). The data shown is for specific binding. Non-specific binding (in the presence of 100 μM GTPγS) was less than 1% of the total binding.

10. Conclusions

Of the functional assays available to study G-protein function, most are extremely useful in allowing the dissection of the G-protein signal transduction pathway. However, it should be noted that the GTP hydrolysis assay, $[\alpha\text{-}^{32}P]$ azidoanilide-GTP and $[^{35}S]GTP\gamma S$ binding assays will only work when examining G-proteins which exchange guanine nucleotides at the same rate as the pertussis toxin-sensitive G-proteins [turnover numbers in the region of 4 min^{-1} (3). For G-proteins with low intrinsic rates of GTP hydrolysis such as G_z (26), it may not be possible to adequately perform the above listed functional assays. Such G-proteins await more sensitive assay techniques.

Acknowledgements

I would like to thank Drs G. Milligan and J. Pouyssegur for providing the facilities to perform the experiments described herein. In addition, I would like to express thanks to Dr W. Rosenthal and Prof. G. Schultz for providing the facilities to utilize $[\alpha\text{-}^{32}P]$ azidoanilide-GTP labelling. I thank EMBO for financial support.

References

1. Rodbell, M., Birnbaumer, L., Pohl, S. L., and Krans, H. M. J. (1971). *J. Biol. Chem.*, **246**, 1877–92.
2. Pfeuffer, T. (1977). *J. Biol. Chem.*, **252**, 7224–34.
3. Gilman, A. G. (1987). *Annu. Rev. Biochem.*, **56**, 615–49.
4. Lochrie, M. A. and Simon, M. I. (1988). *Biochemistry*, **27**, 4957–65.
5. Bourne, H. R., Sanders, D. A., and McCormick, F. (1990). *Nature*, **348**, 125–32.
6. Birnbaumer, L., Abramowitz, J., and Brown, A. M. (1990). *Biochim Biophys. Acta.*, **1031**, 163–224.
7. Ferguson, K. M. Higashijima, T., Smigel, M. D., and Gilman, A. G. (1986). *J. Biol. Chem.*, **261**, 7393–9.
8. Stryer, L. (1988). *Cold Spring Harbor Symp. Quant. Biol.*, **53**, 282–94.
9. Krupinski, J., Coussen, F., Bakalyer, H. A., Tang, W. A., Feinstein, P. G., Orth, K., *et al.* (1989). *Science*, **244**, 1558–64.
10. Litosch, I. (1987). *Life Sci.*, **41**, 251–8.
11. Burch, R. M., Luini, A., and Axelrod, J. (1986). *Proc. Natl. Acad. Sci.*, **83**, 7201–5.
12. Rosenthal, W., Hescheler, J., Trautwein, W., and Schultz, G. (1988). *FASEB J.*, **2**, 2784–90.
13. Brown, A. M. and Birnbaumer, L. (1988). *Am. J. Physiol.*, **254**, 401–10.
14. Fernandez, J. M., Neher, E., and Gomperts, B. D. (1984). *Nature*, **312**, 453–5.

15. Lancet, D. and Pace, U. (1987). *Trends. Biochem. Sci.*, **12**, 63–6.
16. Cantiello, H. F., Patenaud, C. R., Codina, J., Birnbaumer, L., and Ausiello, D. A. (1990). *J. Biol. Chem.*, **265**, 21624–8.
17. McKenzie, F. R., Kelly, E. C. H., Unson, C. G., Spiegel, A. M., and Milligan, G., (1988). *Biochem. J.*, **249**, 653–9.
18. Cotman, L., Herschman, H. and Taylor, D. (1971). *J. Neurobiol.*, **2**, 169–74.
19. Cassel, D. and Selinger, Z. (1976). *Biochim. Biophys. Acta.*, **452**, 538–51.
20. Brandt, D. R. and Ross, E. M. (1986). *J. Biol. Chem.*, **261**, 1656–64.
21. Costa, T. and Herz, A. (1989). *Proc. Natl. Acad. Sci. USA*, **86**, 7321–5.
22. Costa, T., Lang, J., Gless, C., and Herz, A. (1990). *Mol. Pharmacol.*, **37**, 383–94.
23. Gierschik, P., Sidiropoulos, D., and Jacobs, K. H. (1989). *J. Biol. Chem.*, **264**, 21470–3.
24. Horstman, D. A., Brandon, S., Wilson, A. L. Guyer, C. A., Cragoe Jr., E. J., and Limbird, L. E. (1990). *J. Biol. Chem.*, **265**, 21590–5.
25. Higashijima, T., Ferguson, K. M., Smigel, M. D., and Gilman, A. G. (1987). *J. Biol. Chem.*, **262**, 757–61.
26. Casey, P. J., Fong, H. K. W., Simon, M. I., and Gilman, A. G. (1990). *J. Biol. Chem.*, **265**, 21590–5.
26. Laemmli, U.K. (1970). *Nature*, **227**, 680–5.
27. Hames, B. D. (1981). In *Gel electrophoresis of proteins—A practical approach*. (ed. B. D. Hames and D. Rickwood), pp. 1–91. IRL Press Ltd., London and Washington, DC.
28. Hsu, W. H., Rudolph, U., Sanford, J., Bertrand, P., Olate, J., Nelson, C., *et al.* (1990). *J. Biol. Chem.*, **265**, 11220–6.
29. Schleifer, L. S., Garrison, J. C., Sternweis, P. C., Northup, J. K., and Gilman, A. G. (1980). *J. Biol. Chem.*, **255**, 2641–4.
30. Schnefel, S., Profrock, A., Hinsch, K., D., and Schultz, I., (1990). *Biochem. J.*, **269**, 483–8.
31. Goldsmith, P., Backlund Jr., P. S., Rossiter, K., Carter, A., Milligan, G., Unson, C. G., and Spiegel, A. M. (1988). *Biochemistry*, **27**, 7085–90.
32. Gill, D. M. and Meren, R. (1978). *Proc. Natl. Acad. Sci. USA*, **75**, 3050–4.
33. Medynski, D., Sullivan, K., Smith, D., Van Dop, C., Chang, F., Fung, B., *et al.* (1985). *Proc. Natl. Acad. Sci. USA*, **82**, 4311–15.
34. Cassel, D. and Selinger, Z. (1977). *Proc. Natl. Acad. Sci. USA*, **74**, 3307–11.
35. Katada, T. and Ui, M. (1979). *J. Biol. Chem.*, **254**, 469–79.
36. Katada, T. and Ui, M. (1981). *J. Biol. Chem.*, **256**, 8310–17.
37. Burns, D. L., Hewlett, E. L., Moss, J. and Vaughan, M. (1983). *J. Biol. Chem.*, **258**, 1435–8.
38. Kurose, H., Katada, T., Haga, K., Ichiyama, A., and Ui, M. (1986). *J. Biol. Chem.*, **261**, 5423–8.
39. Chang, F. H. and Bourne, H. R. (1989). *J. Biol. Chem.*, **264**, 5352–7.
40. Milligan, G., Unson, C. G., and Wakelam, M. J. O. (1989). *Biochem. J.*, **262**, 643–9.
41. Owens, J. R., Frame, L. T., Ui, M., and Cooper, D. M. F. (1985). *J. Biol. Chem.*, **260**, 15946–52.
42. Tsai, S. C., Adamik, R., Kanaho, Y., Hewlett, E. L., and Moss, J. (1984). *J. Biol. Chem.*, **259**, 15320–3.
43. Pang, I. H. and Sternweis, P. C. (1990). *J. Biol. Chem.*, **265**, 18707–12.

44. Mattera, R., Codina, J., Sekura, R. D., and Birnbaumer, L. (1986). *J. Biol. Chem.*, **261**, 11173–9.
45. Ribeiro-Neto, F., Mattera, R., Grenet, D., Sekura, R. D., Birnbaumer, L., and Field, J. B. (1987). *Mol. Endocrinol.*, **1**, 472–81.
46. Kahn, R. A. and Gilman, A. G. (1984). *J. Biol. Chem.*, **259**, 6228–34.
47. Northup, J. K., Smigel, M. D., and Gilman, A. G. (1982). *J. Biol. Chem.*, **257**, 11416–23.
48. Wong, S. K. F. and Martin, B. R. (1985). *Biochem. J.*, **231**, 39–46.
49. Gordon, J. H. and Rasenick, M. M. (1988). *FEBS Lett.*, **235**, 201–6.
50. Offermanns, S., Schultz, G., and Rosenthal, W. (1991). *Methods Enzymol.*, **195**, 455–75.
51. Schafer, R., Christian, A. L. and Schultz, I. (1988). *Biochim. Biophys. Res. Commun.*, **155**, 1051–9.
52. Offermans, S., Schafer, R., Hoffmann, B. E., Bombien, K., Spicher, K., Hinsch, K. D. *et al.* (1990). *FEBS Lett.*, **260**, 14–18.
53. Pederson, S. E. and Ross, E. M. (1982). *Proc. Natl. Acad. Sci. USA*, **79**, 7228–32.
54. Cerione, R. A., Staniszekski, C., Benovic, J. L., Lefkowitz, R. J., Caron, M. G., Gierschik, P., *et al.* (1985). *J. Biol. Chem.*, **260**, 1493–500.
55. Cerione, R. A., Regan, J. W., Nakata, H., Codina, J., Benovic, J. L., Gierschik, P., *et al.* (1986). *J. Biol. Chem.*, **261**, 3901–9.
56. Asano, T., Pederson, S. E., Scott, C. W., and Ross, E. M. (1984). *Biochemistry*, **23**, 5460–7.
57. May, D. C. and Ross, E. M. (1988). *Biochemistry*, **27**, 4888–93.
58. Smith, C. D., Uhing, R. J., and Snyderman, R. (1987). *J. Biol. Chem.*, **262**, 6121–7.

3

Reconstitution of cyc⁻ membranes by *in vitro* translated G$_s$α: a model for studying functional domains of G$_s$α subunit

YVES AUDIGIER

1. Introduction

In order to investigate the structure–activity relationship on the α subunits of the GTP binding proteins, two main approaches have been designed. These are retroviral transfection of the cDNAs encoding the various α subunits into eukaryotic cells (1) and the synthesis of the α subunits in bacteria which are subsequently used for reconstituting eukaryotic membranes (2, 3). The first approach, although very interesting, is time consuming and the second approach faces the problem of posttranslational modifications which are lacking in the bacterially expressed proteins.

We therefore decided to develop a new strategy which avoids the two disadvantages of the above mentioned approaches: the translation of the α subunit is performed in an eukaryotic system and the *in vitro* translated protein is used directly for reconstituting membranes prepared from a mutant cell line which does not express the α subunit (4). As a corollary, this approach is primarily restricted to the characterization of mutant cells lines that do not express a particular α subunit. Unfortunately, since only G$_s$α subunit satisfies this prerequisite, it introduces a significant limitation to the generalization of this approach.

Nevertheless, the reconstitution can be functionally validated by restoration of the coupling between the receptor and its effector which was deficient in the mutant cell line. Furthermore, since the *in vitro* translated protein is generated from a messenger RNA which comes from *in vitro* transcription of a cDNA, it is possible to analyze the effect of genetic modifications on the different functional properties of the α subunit and therefore to locate the various domains involved in the specific functions which are inherent to the α chain of the GTP binding proteins.

2. Experimental conditions of the *in vitro* reconstitution

2.1 General principles

The ultimate step of the approach is the reconstitution of cyc⁻ membranes (5) (obtained from a mutant S49 cell line which does not express the α subunit of G_s) by the *in vitro* translated α subunit of G_s, the GTP binding protein involved in the activation of adenylate cyclase.

Before reconstitution, the approach begins with the insertion of $G_s\alpha$ cDNA into a transcription vector which then allows the synthesis *in vitro* of the messenger RNA. Translation of this messenger RNA in a cell-free system such as reticulocyte lysate will produce the $G_s\alpha$ protein used for the reconstitution assay. Consequently, besides reconstitution, this approach allows the analysis of the consequences of genetic modifications performed at the nucleotide level on the functional properties of the $G_s\alpha$ subunit at the amino-acid level.

The different steps required for the reconstitution are summarized in *Scheme 1* and they will be described sequentially.

2.2 Subcloning into a transcription vector

In vitro transcription can be accomplished when a commercially available RNA polymerase is added to a cDNA sequence inserted downstream of a promoter for this prokaryotic polymerase. A number of distinct promoters have been used, including the SP6 promoter, the T7 promoter and the T3 promoter, the corresponding RNA polymerase for which has been cloned.

Scheme 2 describes the subcloning of the cDNA in the vector pIBI 31 between the two promoters, T3 and T7. By the alternative use of these two promoters, the sense and the antisense messenger RNAs can be synthesized *in vitro*.

Most of the transcription vectors contain two distinct promoters. However, it should be pointed out that vectors which have the SP6 promoter are not so attractive because of the expensive price of SP6 polymerase.

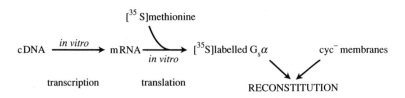

Scheme 1. Different steps of the reconstitution.

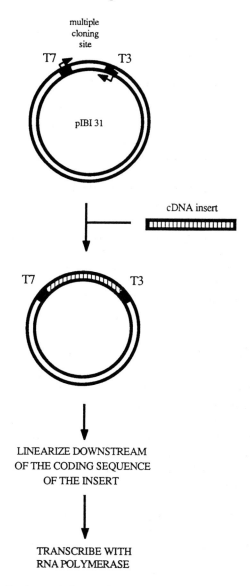

Scheme 2. Subcloning into the transcription vector.

2.3 *In vitro* transcription

Before *in vitro* transcription, the vector has to be linearized at a site downstream of the insert cDNA in order to preferentially transcribe the sequence corresponding to the insert. In contrast to the production of high specific activity RNA probes, *in vitro* transcription has to yield a high amount of messenger RNA: the nucleotide concentration is thus increased to 2 mM and the amount of RNA polymerase is equal to 100 U per microgram of plasmid DNA; [³H]UTP is added as a tracer since it is used only for quantitating the amount of *in vitro* synthesized transcripts (see *Protocol 1*).

Protocol 1. Synthesis of RNA by *in vitro* transcription using T7 RNA polymerase

Materials

- unlabelled NTP mix (12.5 mM each of ATP, CTP, GTP, and UTP in 20 mM Tris–HCl, pH 7.5)
- 1 M dithiothreitol (DTT)
- cap analogue m7 G(5′)ppp(5′)G 5mM (optional)[a]
- 5 × transcription buffer (0.2 M Tris–HCl, pH 8.0, 30 mM MgCl₂, 10 mM spermidine)[b]
- [³H]UTP 20 Ci/mmol at 0.05 μmol/ml
- T7 RNA polymerase
- linearized template DNA

Method

1. Mix the following reagents in a microcentrifuge tube:

unlabelled NTP mix	40 μl
1 M DTT	1.5 μl
cap analogue m7 G(5′)ppp(5′)G 5 mM	20 μl
5 × transcription buffer	50 μl
linearized template DNA (10 μg)	78.5 μl
[³H]UTP 500 pmol	10 μl

2. Pre-incubate 5 min at 37 °C.
3. Add 50 μl T7 RNA polymerase (1000 U).
4. Incubate 1 h at 37 °C.

---

[a] Addition of the cap analog increases the stability of the messenger RNA and is chosen relative to the transcription initiation site of the vector.
[b] Similar reaction conditions are used for T3 RNA polymerase, except that 5× transcription buffer must contain 250 mM NaCl.

After transcription, there are two main procedures for recovering RNA: either the DNA template is removed by digestion with DNAse or the RNA is precipitated by LiCl (see *Protocol 2*).

Protocol 2. Purification of the *in vitro* transcribed RNA

Materials

- SDS 10%
- phenol/chloroform (1/1)
- chloroform/isoamyl alcohol (24/1)
- sodium acetate 3 M
- ethanol
- lithium chloride 10 M
- ammonium acetate 4 M

Method

1. To the 250 μl of transcription medium, add 110 μl water and 40 μl SDS 10%. Vortex.
2. Extract with 400 μl phenol/chloroform. Vortex, centrifuge at 10 000 *g* for 2 min and transfer the upper phase to a new Eppendorf tube.
3. Extract twice with 400 μl chloroform/isoamyl alcohol as described in step 2.
4. Add 45 μl 3 M sodium acetate and 1 ml ethanol.
5. Freeze in liquid nitrogen and then centrifuge at 10 000 *g* for 30 min.
6. Discard supernatant and resuspend the pellet in 120 μl water.
7. Add the same volume (120 μl) of 10 M lithium chloride and leave on ice for at least 5 h.[a]
8. Centrifuge at 10 000 *g* for 30 min.
9. Discard supernatant and resuspend the pellet in 120 μl water.
10. Add 13 μl 4 M ammonium acetate[b] and 330 μl ethanol.
11. Freeze in liquid nitrogen and then centrifuge at 10 000 *g* for 30 min.
12. Discard supernatant. Dry the pellet and resuspend it in 100 μl water.

[a] We routinely perform an overnight incubation.
[b] Ammonium acetate replaces sodium acetate since it can be easily removed by evaporation and therefore will not inhibit further reactions such as *in vitro* translation.

To estimate precisely the amount of RNA that was synthesized, precipitation with trichloroacetic acid is recommended before counting since some unincorporated [3H]UTP may still be present. Knowing exactly the isotopic dilution with cold UTP (in our conditions, dilution factor is 1000) and calculating the number of uridine residues in a transcribed sequence, it is very simple to obtain the number of transcripts and to extrapolate the RNA concentration. From one transcription reaction to the other, there is very little variation and the yield is very close to 1 μg of capped transcript from 1 μg of template DNA. Omitting the cap analog does increase the yield up to four times (4 μg RNA/1 μg DNA), but we have found that it decreases the translation efficiency by the same ratio.

It is important to point out that the transcription reaction can be easily checked by loading 1 μl of the RNA solution (approximately 100 ng RNA) on an agarose gel without denaturation. Under these conditions, the RNA gives a detectable band which migrates close to the dye front following ethidium bromide-staining.

2.4 *In vitro* translation

The RNA is then translated in a cell-free system with a mixture of 19 unlabelled amino acids and a radioactive amino acid, which can be either [³⁵S]methionine, [³⁵S]cysteine, or [³H]leucine (see *Protocol 3*).

Protocol 3. *In vitro* translation

Materials

- reticulocyte lysate
- mixture of 19 amino acids excluding methionine
- [³⁵S]methionine (1000 Ci/mmol)
- purified RNA 100 ng/μl

Method

Nuclease-treated reticulocyte lysate	7 μl
Mixture of 19 amino acids at 1 mM excluding methionine	0.5 μl
[³⁵S]methionine (1000 Ci/mmol; 10 mCi/ml)	1 μl
Purified RNA 100 ng/μl	1 μl

Incubate at 30 °C for 45 min.

We prefer to use the reticulocyte lysate instead of wheat germ extract since the reticulocyte lysate is more likely to perform most post-translational modifications. The choice of the radioactive amino acid depends upon the specific activity of the radioactive amino acid and the occurrence of the same residue in the protein sequence. Although [³⁵S]methionine (1000 Ci/mmol) is the most commonly used amino acid, it is sometimes useful to choose [³H]leucine (120 Ci/mmol) if the number of leucine residues is much larger (more than 10 times) than the number of methionine residues and therefore compensates for the lower specific activity of [³H]leucine.

2.4.1 Titration of the RNA

Like others, we have found that there is an optimal ratio between the RNA dilution and the amount of reticulocyte lysate. Indeed, plotting the amount of translated protein as a function of the RNA dilution gives rise to a bell-shaped curve (see *Figure 1*), suggesting that inhibitors of translation may contaminate the RNA preparation.

Consequently, for each RNA, the titration curve has to be drawn in order to define the RNA dilution which is the most efficient for *in vitro* translation.

i. Analysis of translation products on SDS-polyacrylamide gels

Reticulocyte lysate contains a high concentration of protein (50–60 mg/ml) and therefore it is important to load a small amount of lysate on a SDS-PAGE gel and to dilute the sample in enough sample buffer containing SDS. Using 1.5-mm-thick gels, we have found that 2 μl of translation medium was the maximal volume that can be loaded without producing artefacts in migration of the polypeptides.

Since the apparent molecular weight of $G_s\alpha$ is close to 45 kDa, it is noteworthy to mention the existence of a background [35S]-labelled band which migrates with a mol. wt. close to this value. Fortunately, its mobility can be altered by omitting the normal heating step prior to loading of the SDS-PAGE. Under such conditions this polypeptide migrates with a mol. wt. above 150 kDa (see *Figure 1*).

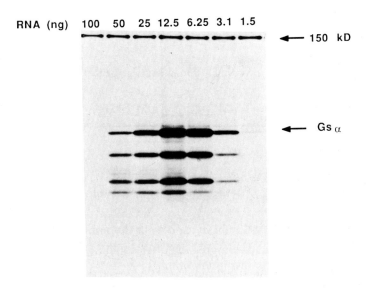

Figure 1. Titration curve of the *in vitro* transcribed RNA

ii. Initiation at internal AUG codons

As reported by many authors, most of the transcription vectors currently used generate, besides the expected translation product, artefactual initiation of the translation at internal AUG condons. Consequently, in addition to the [^{35}S]-labelled band which represents the natural translation product, other [^{35}S]-labelled bands of lower molecular weights are also obtained (see *Figure 1*) which result from initiation at AUG codons which are located downstream in the coding sequence.

2.5 Preparation of plasma membranes from S49 cells

As previously reported (5), the S49 lymphoma cell line and its variant clones are useful for a variety of genetic, biochemical, and pharmacological approaches and therefore techniques have been developed for preparing large amounts of relatively purified plasma membranes.

In order to perform reconstitution between the *in vitro* translated $G_s\alpha$ and the cyc⁻ membranes, we have to prepare the plasma membranes from the mutant cyc⁻ cell line which does not express the α subunit of G_s.

The lymphoma cells are grown in spinner cultures at 37 °C in Dulbecco's modified Eagle's medium containing 10% heat-inactivated horse serum in an atmosphere of 90% air and 10% CO_2.

Protocol 4 is very similar to the procedure described in ref. 6.

Protocol 4. Preparation of plasma membranes from S49 cells

Materials

- PBS buffer: 150 mM NaCl, 5 mM KCl, 1.1 mM KH_2PO_4, 1.08 mM Na_2HPO_4 at pH = 7.2
- cavitation buffer: 150 mM NaCl, 2 mM $MgCl_2$, 1 mM EDTA, 20 mM Na–Hepes at pH = 7.4
- homogenization buffer: 2 mM $MgCl_2$, 1 mM EDTA, 1 mM DTT, 20 mM Na–Hepes at pH = 8.0

Method

1. Harvest cells at a density of 2.0–3.5 × 10⁶/ml.
2. Centrifuge at 200 g for 10 min. Discard supernatant. Wash twice in PBS buffer.
3. Resuspend cells to 3 × 10⁷/ml in ice-cold cavitation buffer.
4. Lyse the cells by rapid decompression after equilibration for 20 min at 28 bar (400 p.s.i.) N_2 in a disruption bomb.
5. Centrifuge the lysate at 900 g for 5 min.
6. Centrifuge the supernatant at 43 000 g for 20 min. Discard supernatant.

7. Resuspend the pellet in homogenization buffer at a final protein concentration of 5 mg/ml.

8. Homogenize in a type B Dounce homogenizer.

9. Store aliquots at −80 °C.

2.6 Reconstitution of cyc⁻ membranes by *in vitro* translated $G_s\alpha$

The basis of the reconstitution step is the addition of the *in vitro* translated α subunit to plasma membranes prepared from the cyc⁻ mutant cell line. As a corollary, the efficiency of the reconstitution has to be related to the membrane-associated fraction and therefore the amount of $G_s\alpha$ which co-sediments with the membrane fraction is a good index of the reconstitution activity.

As shown in *Figure 2*, after centrifugation of the reconstitution medium, there is a significant amount of $G_s\alpha$ which fails to interact stably with the membrane fraction as revealed by the radioactivity found in the supernatant fraction S_1 compared to the membrane fraction P_1. However, the $G_s\alpha$ associated with the reconstituted membranes (P_1 fraction) is not released after subsequent resuspension and centrifugation of the membranes as shown by the localization of the radioactivity quantitatively in the pellet fraction (P_2 fraction).

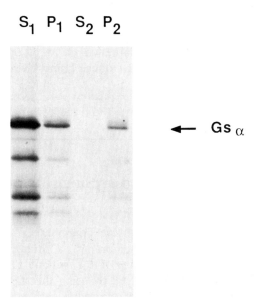

Figure 2. Association of $G_s\alpha$ subunit with the membrane fraction after reconstitution.

Furthermore, we have shown that the characteristics of the association of $G_s\alpha$ in reconstituted cyc⁻ membranes faithfully reproduced those of $G_s\alpha$ in wild-type membranes (4, 7).

Based upon this finding, the influence of several factors such as temperature, the amount of protein or lysate and the concentration of membranes on the amount of $G_s\alpha$ which is recovered in the membrane fraction has been checked. Indeed, there is an optimal ratio between the amount of $G_s\alpha$ and the concentration of membranes. Furthermore, we have found that reconstitution is complete in 30 min at 30 °C or 37 °C.

From these data, we carried out the reconstitution under the conditions described in *Protocol 5*.

Protocol 5. Reconstitution of cyc⁻ membranes by *in vitro* translated $G_s\alpha$

Material
- homogenization buffer: 2 mM $MgCl_2$, 1 mM EDTA, 1 mM DTT, 20 mM Na–Hepes at pH = 8.0

Method
1. Mix 10 µl of translation medium with 10 µl of cyc⁻ membranes (2 mg protein/ml).
2. Incubate at 37 °C for 30 min.[a]
3. Centrifuge the reconstitution medium at 10 000 *g* for 5 min. Discard supernatant.
4. Resuspend the reconstituted membranes in 10 µl of homogenization buffer and homogenize in a type B Dounce homogenizer.

[a] When the reconstituted membranes are prepared to assess generation of cAMP, the incubation is performed at 30 °C in order to decrease inactivation of adenylate cyclase (see Chapter 4).

---

3. Validation of the *in vitro* reconstitution model

3.1 General principles

The properties of the reconstituted cyc⁻ membranes are expected to reproduce those of the wild-type membranes, a situation which can be considered as a biochemical reversion of the genetic defect of the cyc⁻ mutation. This reversion can be assayed by using the standard procedures performed for analysis of the transduction process such as receptor coupling (see Chapter 1) and effector activation (see Chapter 4) but also by designing more specific methods related to the properties of the G_s-mediated pathway such as cholera toxin-catalysed [^{32}P])ADP-ribosylation (see Chapter 2).

3.2 Restoration of the coupling between the *ß*-adrenergic receptor and adenylate cyclase

The phenotype of the reconstituted membranes can be easily determined by measuring the adenylate cyclase activity (see Chapter 4) (8, 9) in response to various agents known to act at different levels of the transduction pathway such as isoproterenol, an agonist of the *ß*-adrenergic receptor, and GTPγS, a hydrolysis-resistant guanine nucleotide which directly activates the α subunit of G_s.

Figure 3 shows the typical phenotype of cyc⁻ membranes characterized by their insensitivity to isoproterenol and GTPγS. After reconstitution by *in vitro* translated $G_s\alpha$, the reconstituted cyc⁻ membranes become sensitive to these agents and thus display the same phenotype as the wild-type membranes.

Restoration of adenylate cyclase activation in the reconstituted cyc⁻ membranes clearly demonstrates that $G_s\alpha$ can correctly interact with the other membrane proteins of the transduction machinery such as the receptor, the *ßγ* subunits and the effector.

It is important to stress that the maximal activation obtained in reconstituted cyc⁻ membranes is less than that observed with wild-type membranes. The

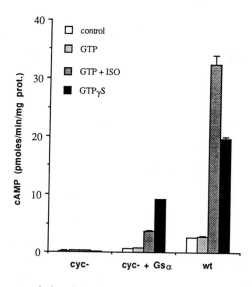

Figure 3. cAMP accumulation induced by various agents on cyc⁻ membranes, re-constituted cyc⁻ membranes and wild-type S49 membranes.

difference between the wild-type membranes and the reconstituted cyc⁻ membranes depends upon the nature of the agent used for activation of the transduction pathway: GTPγS is only twofold less effective in activating adenylate cyclase in reconstituted cyc⁻ membranes whereas isoproterenol is seven times less efficient in the same conditions.

3.3 Stoichiometry of the three components

In this reconstitution model, it is possible to modulate the amount of synthesized protein and therefore analyse the effects of changing the stoichiometry of the various components involved in the transduction process.

As shown in *Figure 4*, when the amount of RNA is increased, there is a concomittant increase in the level of basal and GTPγS-activated adenylate cyclase activity. However, at high amounts of RNA, this activation reaches a plateau, suggesting that the α subunit is no longer the limiting factor in effector activation.

3.4 Restoration of the GTP effect on the affinity of agonist−receptor interaction

A corollary to the effective reconstitution of the transduction pathway is the restoration of the GTP effect on agonist affinity for the β-adrenergic receptor, since it represents the functional consequence of the interaction between the agonist-activated receptor and the GTP binding protein.

In the absence of GTP, the affinity of the agonist for the receptor is high

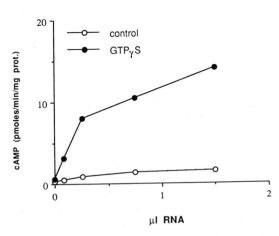

Figure 4. Effect of the $G_s\alpha$ RNA concentration on the basal and GTPγS-stimulated adenylate cyclase activity of the reconstituted cyc⁻ membranes.

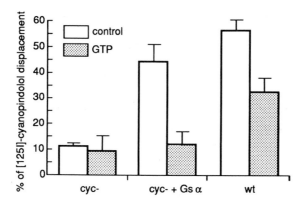

Figure 5. Effect of GTP on the isoproterenol-induced displacement of [^{125}I]cyanopindolol binding on cyc$^-$ membranes, reconstituted cyc$^-$ membranes and wild-type S49 membranes.

whereas it is strongly decreased in the presence of GTP (10) (see Chapter 1). Such an effect is not observed with antagonists and therefore the GTP effect can be measured by analysing the change in the agonist-induced (iso-proterenol) displacement of a radiolabelled antagonist [^{125}I]cyanopindolol) binding in the absence or presence of GTP as shown in *Figure 5* (10).

As expected from the lack of $G_s\alpha$ expression, there is no GTP effect when the binding experiments are performed on cyc$^-$ membranes, i.e. the same concentration (0.1 µM) of isoproterenol displaces the same amount of [^{125}I]cyanopindolol binding in the presence or absence of GTP (*Figure 5*). However, the reconstituted cyc$^-$ membranes do display the GTP effect, since agonist-induced displacement of antagonist-binding is greater in the absence of GTP; this GTP effect is very similar to that observed on wild-type membranes.

3.5 Appearance of a substrate for cholera-toxin-catalysed [^{32}P]ADP-ribosylation

Cholera toxin specifically catalyses ADP-ribosylation of the two main $G_s\alpha$ isoforms which are expressed in wild-type S49 cells (11). Using [^{32}P]NAD, it is thus possible to specifically radiolabel two protein species of approximately 45 Da and 52 Da in wild-type membranes (*Figure 6*).

As expected, no toxin-induced labelling is observed in membranes of the cyc$^-$ S49 variant (11); reconstitution of the cyc$^-$ membranes by *in vitro* translated 52 kDa isoform is accompanied by the appearance of a [^{32}P]-labelled protein which migrates as a 52 kDa species (*Figure 6*).

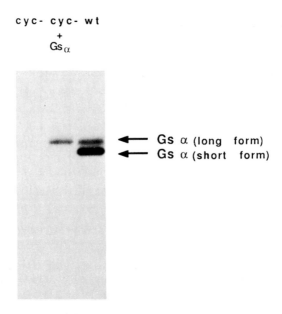

cyc- cyc- wt

+

Gs $_\alpha$

← Gs α (long form)
← Gs α (short form)

Figure 6. Cholera toxin-catalysed [³²P]ADP-ribosylation of cyc⁻ membranes, re-constituted cyc⁻ membranes and wild-type membranes.

4. Applications of the *in vitro* reconstitution model

4.1 General principles

The opportunity of introducing genetic deletions or mutations at the nucleotide level of the cDNA allows the study of functional domains of the α chain of the GTP binding protein, the key element of the transduction machinery. In order to achieve its transduction function, the α subunit has to interact with the receptor, the effector, and the βγ complex. Using this *in vitro* reconstitution model, it is therefore possible to delineate the various peptidic domains which ensure the association with these different protein components.

4.2 Phenotype of a G$_s$α mutant containing a deletion within the amino-terminus

In order to investigate the role of the amino-terminus in the functional properties of the α subunit of G$_s$, we have deleted the nucleotide sequence coding for the 28 amino-terminal amino acids located after the initiating methionine (12). Then, as previously described in *Scheme 1*, the mutated cDNA was *in vitro* transcribed and the corresponding RNA was translated in the reticulocyte lysate system. Finally, the [³⁵S]-labelled mutated α chain which corresponds to $(\Delta 2-29)$ $G_s\alpha$ has been used for reconstituting cyc⁻ membranes.

4.2.1 Coupling of the amino-terminal mutant with the receptor

The cyc⁻ membranes reconstituted by either $G_s\alpha$ or $(\Delta 2-29)$ $G_s\alpha$ have been compared for their ability to display the GTP effect on agonist binding to the β-adrenergic receptor (*Figure 7*).

As shown in *Figure 7*, the amino-terminal deletion considerably decreases the GTP effect although a small but significant difference of agonist affinity is observed when the cyc⁻ membranes are reconstituted by the mutated α chain. These results suggest that the coupling between the β-adrenergic receptor and the mutated α chain is very low.

4.2.2 Adenylate cyclase activity of the membranes reconstituted by the amino-terminal mutant

The basal adenylate cyclase activity and its stimulation in response to GTPγS or isoproterenol have been determined on the cyc⁻ membranes which have been reconstituted by $(\Delta 2-29)$ $G_s\alpha$ (*Figure 8*).

The basal adenylate cyclase activity of the cyc⁻ membranes reconstituted by $(\Delta 2-29)$ $G_s\alpha$ is higher than that of cyc⁻ membranes but much lower than that of the cyc⁻ membranes reconstituted by $G_s\alpha$. Furthermore, corroborating the results obtained with the GTP effect on agonist affinity, the activation of adenylate cyclase induced by isoproterenol on cyc⁻ membranes reconstituted by $(\Delta 2-29)$ $G_s\alpha$ is very low (*Figure 8*). Finally, even with GTPγS which

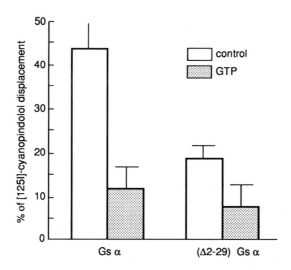

Figure 7. GTP effect on isoproterenol-induced displacement of [¹²⁵I]cyanopindolol binding performed on cyc⁻ membranes reconstituted either by $G_s\alpha$ or $(\Delta 2-29)$ $G_s\alpha$.

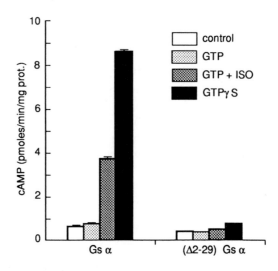

Figure 8. cAMP accumulation induced by various agents on cyc⁻ membranes re-constituted by either $G_s\alpha$ or (Δ2–29) $G_s\alpha$.

directly stimulates the α subunit, there is a very small increase of adenylate cyclase activity.

Consequently, besides impairment of the coupling between the receptor and the transducer, the deletion also modifies the effector activation.

4.2.3 Cholera toxin-catalyzed [³²P]ADP-ribosylation of the amino-terminal mutant

In contrast to $G_s\alpha$, the mutant (Δ2–29) $G_s\alpha$ is not a substrate for [³²P]ADP-ribosylation by cholera toxin although the deletion was not introduced in the sequence containing the arginine residue which is the site for ADP-ribose addition (see Chapter 2). Furthermore, even in the membrane-bound state after reconstitution of the cyc⁻ membranes, there is no detectable [³²P]-labelled band after cholera toxin-catalysed [³²P]ADP-ribosylation.

4.2.4 Interaction of the mutated α chain with $\beta\gamma$ subunits

Sucrose gradient experiments revealed that the mutated α chain cannot interact with the $\beta\gamma$ subunits.

This finding demonstrates that the phenotype of the amino-terminal mutant can be explained by the primary impairment of subunit association which in turn affects $\beta\gamma$ subunits dependent properties such as coupling to the receptor or ADP-ribosylation. However, it remains to be elucidated whether this effect on subunit interaction is also involved in the inability of the mutated chain to activate adenylate cyclase.

5. Other applications

Based on the same principles, our model will allow investigation of the function of other peptide domains in G_s and can also be used for analyzing the activity of chimeraes constructed between the α chains of G_s and other GTP-binding proteins.

Acknowledgements

I should like to thank L. Journot, C. Pantaloni, and J. Bockaert for their various contributions towards the information contained in this chapter.

References

1. Sullivan, K. A., Miller, R. T., Masters, S. B., Beiderman, B., and Bourne, H. R. (1987). *Nature*, **330**, 758.
2. Graziano, M. P., Casey, P. J., and Gilman, A. G. (1987). *J. Biol. Chem.*, **262**, 11375.
3. Mattera, R., Graziano, M. P., Yatani, A., Zhou, Z., Graf, R., Codina, J., *et al.* (1989). *Science*, **243**, 804.
4. Journot, L., Bockaert, J., and Audigier, Y. (1989). *FEBS Lett.*, **251**, 230.
5. Bourne, H. R., Coffino, P., and Tomkins, G. M. (1975). *Science*, **187**, 750.
6. Ross, E. M., Maguire, M. E., Sturgill, T. W., Biltonen, R. L., and Gilman, A. G. (1977). *J. Biol. Chem.*, **252**, 5761.
7. Audigier, Y., Journot, L., Pantaloni, C., and Bockaert, J. (1990). *J. Cell Biol.*, **111**, 1427.
8. Salomon, J., Londos, C., and Rodbell, M. (1974). *Anal. Biochem.*, **58**, 541.
9. Bockaert, J., Deterre, P., Pfister, C, Guillon, G., and Chabre, M. (1985). *EMBO J.*, **4**, 1413.
10. Bourne, H. R., Beiderman, B., Steinberg, F., and Brothers, V. M. (1982). *Mol. Pharmacol.*, **22**, 204.
11. Johnson, G. L., Kaslow, H. R., and Bourne, H. R. (1978). *J. Biol. Chem.*, **253**, 7120.
12. Journot, L., Pantaloni, C., Bockaert, J., and Audigier, Y. (1991). *J. Biol. Chem.*, **266**, 9009.

4

Adenylate cyclase and cAMP

RICHARD W. FARNDALE, LINDA M. ALLAN,
and B. RICHARD MARTIN

1. General introduction

This chapter deals with the methods available to assay adenylate cyclase activity and to determine levels of cAMP in cells. The following provides a brief introduction to place these methods in the context of G-protein action. There are several excellent recent reviews which cover this topic in detail (1, 2, 3).

The involvement of G-proteins in transmembrane signalling was first recognized as a result of studies on adenylate cyclase by Rodbell and co-workers who showed that GTP had a synergistic effect with hormone in activating adenylate cyclase in rat liver membranes. They also were the first to show that analogues of GTP, modified to render the terminal phosphate resistant to hydrolysis, led to a persistently activated state of the enzyme (4). This observation led to the characterization of the GDP/GTP exchange cycle which is thought to operate in all signal-transducing G-proteins.

Hormone-sensitive adenylate cyclase is still by far the best characterized G-protein-mediated signalling mechanism. A number of the hormone receptors involved have been purified and cloned. They appear to have a common general structure consisting of seven transmembrane α helices (see Chapter 1). The G proteins, G_s, which mediates activation, and G_i, which mediates inhibition, were the first to be identified, largely as a result of their sensitivity to ADP ribosylation by cholera toxin or pertussis toxin respectively (see Chapter 2). Together with transducin, the G-protein which mediates rhodopsin action in the eye, the structure and function of G_s and G_i are well characterized. Each is a heterotrimer consisting of α, β, and γ subunits (5, 6). The α subunit contains the sites for interaction with the hormone receptor, catalytic unit of adenylate cyclase or of other effector enzymes, and the guanine nucleotide binding site. It can be said to define the function of the G protein. The β and γ subunits function as a dimer and are thought to control the function of the α subunit. At least in the case of the activation of adenylate cyclase by G_s it appears that the $\beta\gamma$ dimer dissociates from the α subunit on activation. Both cDNA and genomic clones have been obtained for both G_s and G_i (1, 5). There are four forms of G_s produced as a result of differential splicing of the transcript of a single gene. There are three distinct forms of G_i each being the product of a different gene. It seems likely that all forms of G_s activate adenylate cyclase, but it is not certain that all three forms of G_i are able to inhibit. The structure of the α subunits of G_s, G_i, and transducin is quite well characterized. The known structure of a related guanine nucleotide binding protein, the bacterial elongation factor, EFTu was used to assist the generation of a predicted structure based on the amino acid sequences determined from the cDNA clones. The catalytic unit of adenylate cyclase has also been purified and cloned and a structure predicted consisting of ten transmembrane sequences with extensive intracellular domains (7).

The sequence of events in the activation of adenylate cyclase is well characterized. An occupied hormone receptor interacts with the α subunit of G_s leading to its activation. The activated state is stabilized by the binding of GTP to replace bound GDP. The hormone receptor dissociates and is available to interact with other inactive αG_s units, a process which has been called collision coupling. During activation the $\beta\gamma$ dimer is thought to dissociate. The α subunit then directly activates the catalytic unit of adenylate cyclase. It is likely that the catalytic unit and αG_s are persistently associated and that this activation does not require them to move in the plane of the membrane to make contact (8).

2. The binding protein assay for cAMP

The competitive binding assay for cAMP (9) provides a simple and economical means of quantifying the response of tissues, whole cells or cell membrane fractions to agents such as hormones which modulate the activity of adenylate cyclase or cAMP phosphodiesterases. The method relies on the competition between [³H]-labelled cAMP and unlabelled cAMP in the sample for a crude cAMP-binding protein prepared from bovine adrenal glands. Free [³H]cAMP is adsorbed by charcoal and removed by centrifugation, and bound [³H]cAMP in the supernatant is determined by liquid scintillation. The method is sensitive to nM cAMP in the extracted sample. This section of the chapter is intended to detail the preparation of materials and the execution of the assay together with some limitations of the assay and how these may be overcome.

2.1 Preparation of the binding protein

You will need to arrange a source of bovine adrenal glands, such as a local abattoir. The glands should ideally be transported on ice to your laboratory for immediate use, but may be stored frozen until required. The original method calls for extraction of the binding protein (presumably the regulatory subunit of the cAMP-dependent protein kinase) from the adrenal cortex only, the medulla to be discarded. However, a satisfactory binding protein preparation can be made from whole adrenal glands (*Protocol 1*). A recent method (10) for assaying another second messenger molecule, inositol 1,4,5-trisphosphate (IP_3), also employs a binding protein from bovine adrenal glands. *Protocol 2* describes a method of preparing the cAMP-binding protein which is compatible with the preparation of the IP_3-binding protein (see Chapter 5).

Protocol 1. Preparation of the cAMP-binding protein

You will need the following items:

- 5 bovine adrenal glands, on ice
- 250 ml of buffer: 250 mM sucrose, 25 mM KCl, 5 mM $MgCl_2$, 50 mM Tris–HCl, pH 7.4
- nylon mesh (100 μm) or folded muslin
- scalpel or scissors
- Waring blender or tissue homogenizer
- cooled centrifuge capable of 5000 g

1. Trim the fat and connective tissue capsule from the adrenal glands, and weigh them.
2. Chop the glands finely, and blend or homogenize them in 1.5 vol. (based on the weight of the tissue) of ice-cold buffer.
3. Filter the homogenate through the nylon mesh or muslin.
4. Centrifuge the homogenate for 5 min at 2000 g in the cold.
5. Centrifuge the supernatant for 15 min at 5000 g in the cold.
6. Dispense and store the supernatant (the uncalibrated cAMP-binding protein) at −20 °C.

Protocol 2. Alternative preparation of cAMP-binding protein

You will need:

- 12 bovine adrenal glands, on ice
- 500 ml of buffer: 20 mM $NaHCO_3$, 1 mM dithiothreitol, ice-cold
- nylon mesh (100 μm) or folded muslin
- scalpel or scissors
- Waring blender
- large PTFE-on-glass homogenizer

Protocol 2. *Continued*

- cooled preparative centrifuge capable of 35 000 *g*

1. Trim the fat and connective tissue capsule from the glands and cut them open longitudinally.
2. Dissect out and discard the paler medullae.
3. Chop the glands finely and blend for about 2 min with an equal volume of buffer.
4. Strain the suspension through the nylon mesh or muslin.
5. Homogenize the suspension with 20 strokes of the homogenizer.
6. Centrifuge the suspension for 5 min at 5000 *g* in the cold.
7. Centrifuge the supernatant for 20 min at 35 000 *g* in the cold.
8. Store the supernatant (the uncalibrated cAMP-binding protein) at −20 °C.
9. Use the pellet (stored at −20 °C) as the IP_3-binding protein (10).

These protocols yield up to 200 ml of cAMP-binding protein, which should be diluted for use in the assay. The next section (*Protocol 3*) describes the basic assay procedure, which is modified to calibrate the dilution of the binding protein (*Protocol 4*).

2.2 The assay procedure

2.2.1 Construction of the assay

The assay procedure can easily be accommodated within a working day, and it would be feasible to assay in excess of 100 samples at a time.

Protocol 3. Competitive binding assay for cAMP

You will need:

- assay buffer: 4 mM EDTA, 50 mM Tris–HCl, pH 7.4 (Store at −20 °C)
- 32 μM cAMP in water (stable indefinitely at −20 °C)
- [^{3}H]cAMP: 10^6 d.p.m./ml in assay buffer (make up fresh each day)
- binding protein diluted in assay buffer (dilution established using *Protocol 4*)
- charcoal suspensions: 5 mg/ml BSA (for example, Sigma Cat. No. A-4503), 40 mg/ml charcoal (Norit GSX, BDH Ltd). Stable at 4 °C indefinitely
- labelled 1.5-ml polypropylene centrifuge tubes
- 10 000 *g* microcentrifuge at 4 °C

1. Dilute 40 µl of the 32 µM cAMP with 960 µl sample buffer; this dilution contains 64 pmol cAMP per 50 µl.

2. Make serial dilutions of the standard cAMP so prepared by adding, say, 200 µl standard cAMP to 200 µl sample buffer, mixing well and repeating this procedure until 125 fmol cAMP per 50 µl is reached. You should have a range of standards of 64, 32, 16, 8, 4, 2, 1, 0.5, 0.25, 0.125, and finally, 0 pmol cAMP per 50 µl in sample buffer.

3. Pipette 50 µl of these (in duplicate) into labelled centrifuge tubes.

4. Add 50 µl of [^{3}H]cAMP; mix well.

5. Add 100 µl of dilute binding protein.

6. Prepare tubes similarly containing 50 µl of your unknown samples, 50 µl of [^{3}H]cAMP and 100 µl of dilute binding protein as above.

7. In addition, prepare two tubes (the charcoal blanks) containing 50 µl of sample buffer, 50 µl of [^{3}H]cAMP and 100 µl of assay buffer in place of binding protein.

8. Mix all tubes well and place on ice, or in the cold for 2–3 h.

9. At the end of this time, open a batch of tubes, add 100 µl of the charcoal suspension, mix well and centrifuge for 2 min in the cold at 10 000 g.

10. Remove a 200 µl sample of the supernatant to a scintillation vial, add scintillant and count ^{3}H.

11. Count also a 50 µl sample of the [^{3}H]cAMP (the total count in the assay).

The composition of the sample buffer in which the standards are diluted will vary with the extraction technique used; it may be adequate to use assay buffer, or it may be that samples have been prepared in tissue culture medium, in which case the standards should all be diluted in the same medium. This should also contain all general additives, such as phosphodiesterase inhibitors or solvents for hormones, at the same concentration as used in your experiment.

It is important that the sample and [^{3}H]cAMP should be well mixed before the binding protein is added, otherwise the binding protein may adsorb most of the [^{3}H]cAMP, which will take considerable time to re-equilibrate on thorough mixing with the sample. Do not be tempted to save a pipetting step by mixing the [^{3}H]cAMP with the binding protein before dispensing them!

2.2.2 Calculation of results

Subtract the counts obtained from the charcoal blanks from all other values. (The charcoal blank corresponds to that portion of the [^{3}H]cAMP not removed by the charcoal in the absence of binding protein.) Next, express the counts for samples or standards as a percentage of the toal count. Next, on semilogarithmic paper, plot the percentage counts for each standard against its cAMP content (*Figure 1*). Determine the cAMP content of the unknown samples by interpolation from this curve.

This method offers an absolute check on the performance of the assay, since the percentage recovery of [³H]cAMP in the absence of added cAMP (the zero standard) is determined for each run of the assay. Hence if the affinity of the binding protein is compromised for any reason, this is immediately apparent. Also, many scintillation counters offer a programme specifically for radio-immunoassay which performs the calculation described above.

2.3 Calibration and storage of the binding protein

The preparation procedures for the binding protein yields material containing typically 25 mg protein/ml. The best working dilution (typically tenfold) should be determined for each preparation by setting up assay tubes as above each containing 2 pmol cAMP per 50 µl, [³H]cAMP as above and 100 µl of binding protein as prepared or diluted by from 2–20 times with assay buffer. The dilution which gives maximum recovery of [³H]cAMP, typically about 20% of total counts, is that which should be used routinely. The binding protein should then be dispensed in appropriate portions, and stored at −20 °C or lower. Under these conditions it is stable in our hands for at least six years.

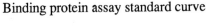

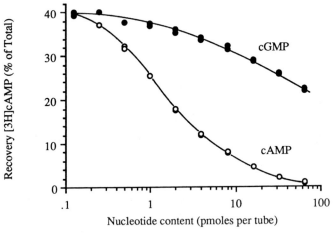

Figure 1. A standard curve for the binding protein assay obtained using duplicate samples of cAMP (○) or cGMP (●) to compete with the radiolabel. The binding protein was prepared by *Protocol 2*.

Thaw and dilute the binding protein for use immediately before the assay; its binding affinity is liable to decline if the working dilution is stored at 4 °C for more than 24 h or so.

2.4 Factors affecting the assay

2.4.1 Timing of the assay procedure

The interaction between binding protein, cAMP and [³H]cAMP is near equilibrium by 2 h, but changes rapidly during the first few minutes of incubation (see *Figure 2*). It is important that all samples have had roughly equal incubations, therefore, especially when critical comparisons of similar [cAMP] are to be made.

Similarly, the apparent recovery of [³H]cAMP after the addition of charcoal varies linearly with time (see *Figure 3*). This problem has been addressed by using saturated $(NH_4)_2SO_4$ to precipitate the binding protein without displacing the complexed cAMP from it (11), hence requiring the free [³H]cAMP in the supernatant to be collected and counted. In our hands, this procedure is liable to greater variability than the charcoal method, and so we prefer to control the time of the assay procedure instead. This can be achieved by minimizing the batch size to be centrifuged at a time, and by randomizing sample groups which are to be compared.

These precautions may be unnecessary if the experimental changes in [cAMP] are large; coefficient of variation of around 5% can readily be obtained even from small groups of, say, four samples.

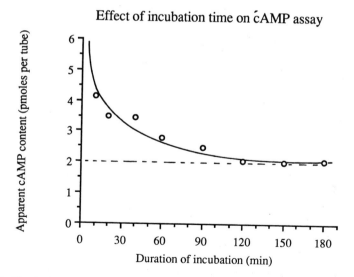

Figure 2. The importance of allowing adequate equilibration time after adding binding protein to the assay tube, and before adding charcoal.

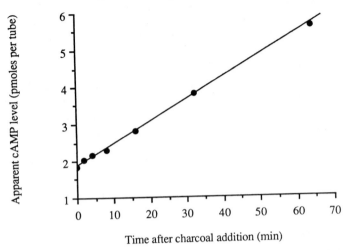

Figure 3. The apparent increase in [cAMP] with increase in time after adding charcoal and before centrifugation. These values were obtained by comparison with a 2-min delay.

2.4.2 Interference in the assay

The extracted samples may contain substances which potentially might inhibit cAMP binding to the protein. It is important for absolute [cAMP] determinations that the standard curves be constructed in the same medium as the samples, since a standard curve constructed in saline may be somewhat lower than if water or assay buffer were used. This is straightforward for samples prepared using boiling or ethanol/HCl extraction, where the boiled suspending medium or the assay buffer are used. For perchloric acid (PCA) extracts, neutralised PCA/sample medium should be used as diluent for the standard curve.

Other additions to the medium, such as ethanol or DMSO (0.1%) are not highly active interferants but should be considered if higher concentrations are used. Similarly, other nucleotides extracted from the sample do not in our hands interfere: ATP, ADP, AMP, or p[NH]ppA (an alternative substrate for adenylate cyclase) at 1 mM and NAD at 0.5 mM are inactive. Others have reported interference from ATP whilst using the assay to measure adenylate cyclase activity in isolated membrane preparations (12). If this should prove a problem, the ATP can be removed from the sample by adding neutral alumina, to which it binds specifically, then centrifuging and assaying cAMP in the supernatant (see *Protocol 9*, below).

cGMP, however, does interfere with the assay, and has about two orders of magnitude lower affinity for the binding protein than cAMP (*Figure 1*), so that it may be important to consider its possible effects when interpreting results. The effect of cGMP might be significant at threshold levels of cAMP, but 50 pmol of cGMP added to 2 pmol cAMP in our hands caused slight overestimation by about 0.2 pmol.

The structural similarities of the xanthines which enable them to serve as phosphodiesterase inhibitors do not generally lead to cross-reaction with the cAMP-binding protein. We have suspected interaction between the xanthines and some culture media, however (especially Grace's insect cell medium), and so it is important that such potential positive interference be controlled, by making up the standards in the same medium as the samples.

The assay is sensitive to pH. This is a particular complication of the acid cAMP extraction procedures (see Section 2.7), so it is essential that the extracted samples are neutralized, so that their pH lies in the range 6.5 to 8.0.

2.5 Sensitivity and range of the assay

Figure 1 shows that the semilog standard curve is roughly linear from 250 fmol to 8 pmol per 50 µl (i.e. 5 to 160 nM cAMP), and may be readable outside this range. The assay procedure detailed above includes standards of 32 and 64 pmol per 50 µl, which may be useful to indicate necessary sample dilution rather than to be read accurately. The sensitivity may be limited by the specific activity of the [^{3}H]cAMP used. These results were obtained with material of 41 Ci/mmol (Amersham TRK.498), and a slight increase in sensitivity could potentially be obtained with higher specific activity [^{3}H]cAMP. Note that the cAMP associated with the radiolabel amounts to about 600 fmoles per assay tube. Substituting the [^{125}I]succinyltyrosine methyl ester of cAMP (the radiolabel used for radio-immunoassay, Amersham IM.106, see below) is not a viable means of increasing sensitivity since the binding protein has a low affinity for this material.

2.6 Extraction of cAMP from experimental materials

Criteria for a useful extraction method are that the adenylate cyclase and phosphodiesterase activities of the sample should cease promptly; that cAMP be extracted quantitatively or reproducibly from the sample, and that cAMP can be transferred to the assay procedure at useful concentration and in a medium which does not compromise the cAMP:binding protein interaction. The method of choice depends on the nature of the sample. For cells in suspension, it may be adequate to stop the experiment by boiling briefly; for cells in monolayer culture or for intact tissues a more complicated method is called for. Three useful methods are described below. Each of these, of course, requires adequate [cAMP] in the sample for subsequent detection in the assay. The first would be the method of choice for cells in suspension.

Protocol 4. cAMP extraction from cell suspensions by boiling

You will need:

- 1.5-ml polypropylene centrifuge tubes
- water bath at 95–100 °C fitted with racks to take centrifuge tubes
- 10 000 g microcentrifuge

1. Incubate cell suspension with hormones, phosphodiesterase inhibitors, etc.
2. After appropriate incubations, place samples in centrifuge tubes in the boiling water bath.
3. Remove after about 5 min, mix well, centrifuge for 2 min.
4. Take 50 µl samples for assay.

We have used this method successfully for insect gut cells in suspension culture at 10^6 cells/ml in Grace's medium (13), for lymphocyte cell lines at 5×10^7 cells/ml in RPMI medium (14), for freshly isolated hepatocytes at 10^6 cells/ml in Krebs–Ringer buffer, or for platelets at 10^9/ml in Tyrrode's saline (unpublished data). The chief advantage of this method is brevity and simplicity. It may be complicated by the presence of serum or BSA in the medium, which may gel on boiling, so some adjustment of protein concentration (to less than 5 g/litre) may be necessary. It may also be possible to concentrate cells in suspension prior to experiment to achieve detectable levels of cAMP, although this should be done with caution since cellular ATP levels may be compromised if anaerobic conditions result from this procedure. Another approach if cAMP levels are low might be to use phosphodiesterase inhibitors such as 100 µM isobutylmethylxanthine (IBMX); indeed it may in any case be sensible to test cells with and without IBMX since elevated cAMP is unlikely to persist once the phosphodiesterase begins to degrade it.

Protocol 5. cAMP extraction from cell culture dishes

You will need:

- ethanol containing 17 ml conc. HCl per litre, on ice
- freezer (−20 °C)
- polystyrene or glass tubes of about 5 ml capacity
- vacuum line to remove culture medium
- vacuum desiccator containing NaOH pellets
- vacuum pump to evaporate the ethanol/HCl (with suitable protective trap)

1. Incubate cells (in 30 mm cell culture dishes) for appropriate times with effector materials.
2. Remove medium with vacuum line, and add 1 ml ice-cold ethanol/HCl.
3. Transfer initially dish to ice, then to the freezer overnight.
4. Transfer the ethanol/HCl to polystyrene tubes and place in desiccator.
5. Carefully evacuate to about 0.5 atmospheres, then 30 min later to full vacuum.
6. When completely dry (perhaps after 2 days or so) reconstitute the sample in about 100 µl of assay buffer. Check that this is not acidic, and assay 50 µl as *Protocol 3*.

Points to bear in mind when designing such assays are that adenylate cyclase activity might be sensitive to pH and to temperature, both of which might be perturbed on removing dishes from their tissue culture incubator. Perturbation through handling of cultures might be sufficient to elevate cAMP levels as a result of endogenous prostaglandin synthesis in the samples under study (15, 16). For these reasons we often set up such incubations using Hepes or Mops rather than bicarbonate-buffered media, maintaining temperature by placing the dishes on an aluminium plate just below the surface of a 37 °C water bath, and allowing 15 min or so to elapse after changing the medium before starting the exposure to experimental ligands. These are added to the tissue culture medium in a negligible volume, and mixed by gentle swirling.

It is important that the extracted and reconstituted sample should be neutral, since the assay is pH-sensitive. It may be necessary to pump the desiccator for extended times to achieve this, or to leave the samples evacuated over NaOH for several days. cAMP appears stable under these conditions. The method also allows for some concentration or dilution of the sample relative to its tissue culture conditions, which may be a useful feature. We have applied this method, which would be our first choice, to fibroblasts and osteogenic cells in confluent cultures (15, 17). An alternative extraction which may be more efficient, but is more tedious and in our hands less reproducible, is given below. This method can be adapted to whole tissues, cell monolayers or suspensions.

Protocol 6. cAMP extraction using perchloric acid

You will need:

- 3 M perchloric acid in water (200 ml per litre)
- 2 M NH_4OH in water
- 1.5-ml polypropylene centrifuge tubes

Protocol 6. *Continued*

- microcentrifuge
- vacuum line to remove culture medium
- pH papers

A. *Cell suspensions*

1. Incubate 200 μl cell suspension (in centrifuge tubes) with the effector materials.

2. Stop at appropriate times by adding 40 μl 3 M perchloric acid. Mix well and centrifuge the protein pellet. Remove the supernatant to another centrifuge tube.

3. Neutralize by adding 2 M NH$_4$OH dropwise.

4. Leave on ice for 2 h for the perchlorate to precipitate.

5. Centrifuge, check the supernatant is neutral, and take 50 μl for assay by *Protocol 3*.

B. *Tissues*

1. Incubate the tissues with the effector materials.

2. Stop by transferring the tissue to 0.5 M perchloric acid, prepared by diluting stock in water.

3. Mix well, centrifuge, remove the supernatant and neutralise it, and then proceed as above.

C. *Cell monolayers in culture*

1. Incubate the dishes with the effector materials.

2. Stop the reaction by removing the culture medium and replacing it with 1 ml 0.5 M perchloric acid.

3. Transfer to centrifuge tube, and proceed as above.

These methods have the advantage that they remove proteins from the sample, and the rigorous conditions should provide efficient extraction of cAMP from difficult materials such as intact tissues. Against this, the neutralization procedure is tedious, and whilst trials with blank samples will indicate how much NH$_4$OH to add to each tube, it is important that they should be checked for neutrality individually. This might be assisted by adding indicator to each tube (10).

2.7 Alternative assays: radio-immunoassay for cAMP

RIA for cAMP usually employs an antiserum raised against the succinyl-tyrosine methyl ester of cAMP; as it stands, this method is rather more sensitive than the above (typically 20 fmol–2 pmol per assay tube). However, the antiserum has a much higher affinity for succinylated or acetylated cAMP, and with this modification, a few femtomoles of cAMP can be detected per assay tube (18). Both the antiserum (NEN du Pont NEK-034; Miles-Yeda) and the radiotracer [^{125}I]succinyltyrosine methyl ester of cAMP; Amersham IM106, NEN du Pont NEX-130) are commercially available separately or in kit form, and the suppliers provide protocols for the acetylation or succinylation procedure. These methods provide high sensitivity for the detection of cAMP, although they are more tedious and the consumable costs are very substantially greater than the cAMP-binding protein method.

3. Determination of adenylate cyclase activity

3.1 Introduction

Adenylate cyclase activity is usually determined in plasma membrane preparations which range in purity from quite highly purified material, where gradient centrifugation has been employed, to crude particulate fractions from a cell homogenate. There are two widely used methods. In the first, the binding protein assay described above is used to determine the amount of cAMP present in a sample after a timed incubation. The second method devised by Salomon *et al* (19) uses [α^{32}P]ATP to generate [^{32}P]cAMP followed by a simple two-column chromatography system to separate labelled cAMP from other labelled adenine nucleotides. The two approaches have different advantages and disadvantages.

3.2 The binding protein method

The binding protein approach has the advantage of lower cost, largely because the cost of the radio isotopes required is substantially less. It is less hazardous, depending upon [^{3}H] rather than [^{32}P] labelling. Finally, it does not require any special apparatus beyond the basic equipment found in most laboratories. There are two main disadvantages. First, the change of radioactivity is not linear with cAMP concentration so that as the concentration of cAMP increases, the accuracy declines (see *Figure 1*). This is not usually a problem when measuring concentrations of cAMP in cell extracts since it is rare for the concentration to vary by much more than an order of magnitude. However, in an adenylate cyclase assay using a plasma membrane preparation, the activity may vary by a much greater factor so that the loss of accuracy as higher concentrations of cAMP are achieved may become a problem, to the extent that the determination needs to be repeated at a different dilution. The second disadvantage of the binding protein approach is that it is much more tedious than the Salomon method (19) to the point where it becomes impractical if several hundred determinations are required in a week. The [α^{32}P]ATP method is then the method of choice. However, it requires specialised equipment, so if a small number of determinations over a short period are required the binding protein assay will be adequate. A second reason for preferring the binding protein assay is if the substrate to be used is p[NH]ppA rather than ATP since [α^{32}P]-p[NH]ppA is much less readily available then [α^{32}P]-ATP. Finally, if a high concentration of ATP is required in the assay, in excess of 1 mM, the binding protein method may be preferred since the cost of the labelled ATP required to maintain a sufficiently high specific activity in the assay may be prohibitive.

Protocol 7. Adenylate cyclase determination by binding protein method

You will need:

- 1.5-ml polypropylene centrifuge tubes
- water bath at 95–100 °C fitted with racks to take centrifuge tubes
- water bath at typically 30 °C
- 10 000 g microcentrifuge
- membrane preparation (typically 1 mg protein/ml in 1 mM dithiothreitol, 25 mM Tris–HCl, pH 7.4, held on ice)
- assay mix: ATP (typically 500 μM), MgCl$_2$ (typically 10 mM) creatine phosphate (10 mM), creatine kinase (50 Units/ml), theophylline (1 mM) or IBMX (100 μM), dithiothreitol (1 mM) in Tris–HCl buffer, pH typically 7.4. The concentration of ATP and MgCl$_2$ and the pH may be varied to suit the needs of the experiment.

1. Warm the centrifuge tubes containing the assay mix and any other additions (such as hormones), in a volume of typically 60 μl, to 30 °C.
2. Start the incubation by adding, typically, 40 μl of membrane suspension to the centrifuge tubes.
3. After appropriate times, place the samples in centrifuge tubes in the boiling water bath.
4. Remove after about 5 min, mix well, centrifuge for 2 min.
4. Take 50 μl samples for competitive binding assay as in *Protocol 3*.

The creatine phosphate and creatine kinase serve as a regenerating system to maintain the ATP. Theophylline or IBMX are required to inhibit breakdown of cAMP by phosphodiesterases in the plasma membrane preparation.

Some workers have reported that ATP interferes with the binding of cAMP to the binding protein and impairs the sensitivity of the assay (12), and an adenylate cyclase assay mix usually contains much more ATP than the boiled extract from a cell preparation. In our laboratory we do not observe any effect of ATP or any other adenine nucleotide at concentrations up to 1 mM in the assay mix. The difference in our experience from other laboratories probably reflects either differences in the individual binding protein preparation or that other groups may have used higher concentrations of ATP. Our practice is to determine the level of cAMP in a 50 µl sample from the stopped incubation without further treatment. However, if a binding protein preparation is to be used for the assay of adenylate cyclase rather than for determining cAMP levels in cell preparations it is essential to check whether the standard binding curve is affected by the highest concentration of ATP to be used. If it is, ATP might be removed by adsorption to neutral alumina according to the *Protocol 8* (12).

Protocol 8. Treatment of stopped assay to remove ATP

You will need a stirred slurry of neutral alumina consisting of 0.45 g/ml in 50 mM triethanolamine/HCl buffer pH 7.4.

1. Add an equal volume of the slurry (typically 100 µl) to the boiled assay mix.

2. Centrifuge in a microfuge to sediment the alumina.

3. Take 50 µl samples for competitive binding assay as in *Protocol 3*.

The addition of alumina should not remove any of the cAMP. This should be checked by monitoring the effect of alumina addition on $[^3H]cAMP$ in a standard sample. Alternatively, if the effect of ATP on the binding protein assay is found to be small it is preferable and more convenient to avoid this procedure by including the appropriate level of ATP in the standard curve.

3.3 The $[\alpha^{32}P]ATP$ method

3.3.1 General principles

The assay employs a two column system to separate $[^{32}P]cAMP$ from labelled ATP, ADP, AMP, and P_i. The first column is Dowex 50 which being negatively-charged would not be expected to bind any of these components. However, despite being acidic, cAMP does adsorb to the column, probably by a non-specific interaction with the resin base allowing the other labelled components to be washed away. The cAMP is then washed onto a column of neutral alumina to which it adsorbs together with any remaining ATP. Alumina binds cAMP less avidly than other adenine nucleotides since the cyclization leads to the loss of the vicinal hydroxyls on the ribose ring. Cyclic AMP can be washed off with imidazole buffer. Using this method, non-specific counts vary from zero to about 1 count for every 10^6 added as $[\alpha^{32}P]$-ATP.

3.3.2 Equipment

The arrangement of the equipment required is shown in *Figure 4*. The assay requires two sets of columns, one for the Dowex 50 and one for the alumina. The bed volume of the columns is approximately 1 ml and it is necessary to accommodate a volume of at least 10 ml above the packed material. Any column which satisfies these requirements can be used. In our laboratory we use glass columns made by a local company with the dimensions shown in *Figure 4*. A plug of glass wool is inserted into the bottom of the column. Disposable plastic columns are also satisfactory. Columns with a total volume of 10 ml supplied by Bio-Rad (Catalogue No. 731–1550) can be used but the plastic sinter supplied needs to be replaced with glass wool to give a satisfactory flow rate. Racks for the columns should be made of Perspex since this is easy to clean from radioactive contamination and provides protection from [^{32}P] radiation. They will need to be purpose built to fit the other equipment available. For each set of assays two racks are required, one for the Dowex columns and one for the alumina columns. The Dowex column rack must locate on top of the alumina column rack so that the columns will drip into the tops of the alumina columns. The alumina columns in turn need to drip into scintillation vials. Modern scintillants such as Zinsser Quicksafe A, which we use, will accommodate the scintillant and the eluate from the alumina column in a 5-ml mini vial. The most convenient arrangement is to design the racks such that the alumina columns drip into vials already in the racks which will go into the scintillation counter. For example, in our laboratory we use a Beckman LS 3801 which accommodates 648 mini-vials in racks containing 18 vials each. Our column racks are therefore designed to take 6 rows of 18 columns, 108 in all, so that for most experiments one set of columns is sufficient.

Protocol 9. Preparation of columns

You will need:

- a stirred slurry (50% v/v) of Dowex 50 ×4 200 in 1 M HCl
- dry neutral alumina (Sigma type WN3, Catalogue No. A9003)
- plastic or glass columns plugged with glass wool

1. Dowex columns: using a 5-ml Gilson Pipetman or similar pipetter, pipette 2 ml of the slurry into each column. This gives a bed volume of 1 ml.
2. Alumina columns: weigh 0.8 g of alumina into a small glass tube. Then mark the tube wall to calibrate the volume. Tip the alumina into the column. The tube can then be used to measure the amount for the rest of the columns. This is sufficiently accurate and very much less tedious than weighing for each individual column.

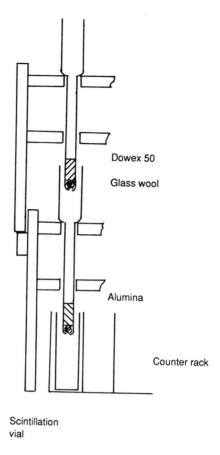

Figure 4. The arrangement of the columns and scintillation vials for the $[\alpha^{32}P]$-ATP method. The dimensions of the glass columns used in our laboratory are as follows: *top section* 5 cm/2 cm; *bottom section* 10 cm/1 cm.

3.3.3 The assay procedure

The assays are conducted in a similar fashion to *Protocol 8*, but substituting a different tube and including $[\alpha^{32}P]$ATP. The concentration of ATP and $MgCl_2$ and the pH may be varied to suit the needs of the experiment.

We have found the assay to be applicable to a very wide range of tissue preparations, including trypanosomes (20), insect gut cells (13), and *Escherichia coli* (21), as well as plasma membrane preparations from many mammalian tissues such as liver (8) and platelet (22). This method has been applied to hundreds of different membrane preparations as well as the reconstituted proteins of the adenylate cyclase complex.

Protocol 10. Adenylate cyclase assay using [α^{32}P]ATP

You will need:

- suitable assay tubes, (for example, LP3 polystyrene tubes: Luckham Ltd, Victoria Gardens, Burgess Hill, Sussex)
- water bath at typically 30 °C
- membrane preparation (typically 1–2 mg protein/ml in 1 mM dithiothreitol, 25 mM Tris–HCl, pH 7.4) held on ice
- basic assay mix (with final concentrations given): ATP (typically 100 μM), 10^6 c.p.m. [α^{32}P]ATP, MgCl$_2$, (typically 10 mM) 500 μM cAMP, 10 mM creatine phosphate, creatine kinase (50 Units/ml), 1 mM dithiothreitol in 25 mM Tris–HCl buffer, pH typically 7.4. Retain a sample for counting (see *Protocol 11*)
- stop solution: 2% SDS, 40 mM ATP, 1.4 mM cAMP (can be stored frozen)
- [^{3}H]cAMP recovery standard: 0.25 μCi/ml in water. Retain a sample for counting.

1. Dispense the assay mix and any other additions into LP3 tubes; transfer to water bath.
2. Start the reaction by adding, typically, about 50 μg membrane protein in 40 μl. Mix well.
3. After suitable times (less than 30 min) add 100 μl of stop solution. Mix well.
4. Add 100 μl [^{3}H]cAMP recovery standard. Mix well.
5. Add 700 μl water. Samples may be frozen at this point for subsequent analysis.

The [α^{32}P]ATP level in the assay may be increased if greater sensitivity is required or reduced if the preparation has a particularly high adenylate cyclase activity. The cAMP is required to trap the radioactive cAMP produced and to protect it against hydrolysis by phosphodiesterase. Cyclic AMP is a very weak competitive inhibitor with ATP and should not exceed the ATP concentration by more than a factor of 5. If a low concentration of ATP is to be used, the concentration of cAMP may need to be reduced. The creatine phosphate and creatine kinase serve as a regenerating system to maintain the ATP. The creatine phosphate concentration may need to be increased if the ATPase activity in the preparation is high. The ATP used should be of the highest purity available, a particular problem being contamination with GTP. Sigma Grade I ATP (Catalogue No. A2383) is prepared by chemical phosphorylation of adenosine and is low in contamination with GTP.

The total volume of the assay is usually 100 μl. It is convenient to add the basic assay mix in a volume of 40 μl, to use 20 μl for various other additions such as guanine nucleotides or hormones, and to initiate the incubation by adding the membrane preparation in a total volume of 40 μl. The assay is highly reproducible and for most purposes triplicate samples will be quite adequate. Triplicate zero time samples are also required consisting of the assay mix, stop solution and 40 μl of water in place of the membrane preparation. The incubations are performed for a fixed time at appropriate temperature, for example 30 °C. It should be noted that the catalytic subunit of adenylate cyclase is not stable at physiological temperature in many membrane preparations. For most purposes, therefore, we prefer to hold the membrane suspension on ice and pipette cold portions into the warmed assay mix to start the reaction. This necessarily means that the activation profile includes a component due to the warming of the cold membranes. In most cases this is not significant, but for detailed examination of the activation kinetics it may be essential to pre-warm the membranes as well as the assay mix before starting the reaction.

The purpose of the SDS is to dissolve the membranes and to stop the activity. The ATP dilutes the radiolabelled ATP in the assay by about 100-fold and further ensures an effective stopping of the reaction. Cyclic AMP is needed as a carrier for the following separation step. It does not have anything to do with stopping the reaction but it is convenient to add it at this point. The ATP used here can be relatively crude, such as Sigma Grade II (Catalogue No. A3377). The [^{3}H]cAMP recovery standard gives approximately 15000 c.p.m. in 100 μl, which is added to each of the samples to enable the yield of [^{32}P]cAMP to be corrected for losses during the column separation.

Protocol 11. Separation of [^{32}P]cAMP after adenylate cyclase assay

You will need:

- variable dispensers containing (a) water and (b) 0.3 M imidazole, pH 7.4
- racks of acid washed Dowex 50 columns (see step 5, below) primed with 15 ml water
- racks of alumina columns, primed by washing with 10 ml 0.3 M imidazole, pH 7.4
- racks of scintillation vials to receive the samples
- suitable scintillant; for example, Quicksafe A (Zinnser Analytic, Howarth Rd, Maidenhead, SL6 1AP)

1. Decant the sample from *Protocol 10* on to the Dowex columns and allow to run to waste.
2. Add 1 ml of water and allow to run to waste. Repeat this step.

Protocol 11. *Continued*

3. Locate the Dowex columns on top of the alumina columns.

4. Add 6 ml of water to the Dowex columns and allow to run through the alumina columns to waste.

5. Remove the Dowex columns and regenerate by washing with 3 ml of 1 M HCl.

6. Add 1 ml of 0.3 M imidazole buffer to the alumina columns and allow to run to waste.

7. Locate the alumina columns on top of the scintillation vials in their rack.

8. Add 1.75 ml of 0.3 M imidazole buffer to each column and allow it to run into the scintillation vials.

9. Remove the alumina columns.

10. Add 3 ml of scintillant to each scintillation vial, cap and mix thoroughly—for example, by inversion ten times.

11. Count the vials together with the standard vials (see below).

Priming the columns can be done at any convenient time on the day of the experiment as can regenerating the Dowex columns with HCl. When the columns are set up initially, it is worth checking the elution profile using labelled ATP and labelled cAMP since it may vary slightly with different column geometries. Minor adjustments can then be made to the volumes above.

i. Standard vials

These should be prepared in duplicate. The relevant materials are added to a vial containing scintillant, and 1.75 ml of 0.3 M imidazole buffer. The standards are:

(a) backgrounds containing imidazole only.

(b) the [^{3}H]cAMP recovery standard (100 μl of the solution used in *Protocol 10*)

(c) the [α^{32}P]ATP standard. Dilute the assay mix 100-fold, i.e. 10 μl into 1 ml. Take the same volume added to each assay tube in *Protocol 10*; for example, 40 μl. This will give a reasonable number of counts in the vial, about 10 000 c.p.m., and will be used to determine the total number of counts in the assay.

3.3.4 Scintillation counting and calculation of results

The vials are counted with an appropriate dual label programme for ^{32}P and ^{3}H. The counting channels are adjusted so that there is no spillover of counts from the ^{3}H channel into the ^{32}P channel. Since all samples are counted under identical conditions of quenching, we use c.p.m. rather than d.p.m. in the calculations to avoid approximations introduced by quench correction. However, it is important that the quench parameters provided by the instrument are monitored to identify occasional problems caused by dispensing the wrong volume of either imidazole eluant or scintillant. It is also important that the samples and scintillant are very well mixed; the volumes of fluids given here lead to the formation of a stable gel rather than a clear emulsion. Inadequate mixing is best identified from the $[\alpha^{32}P]$ATP standard c.p.m., which will show increased counts in the ^{3}H channel (due to Cerenkov radiation, which is of low energy) and reduced counts in the ^{32}P channel when the $[\alpha^{32}P]$ATP is not intimately mixed with the scintillant.

The protein concentration in the membrane preparation is also required so that the protein content of each assay is known.

Protocol 12. Calculation of results

1. Subtract the background c.p.m. for ^{3}H and ^{32}P respectively from all standard and sample c.p.m. Some counters will do this automatically.

2. Calculate the specific activity of $[\alpha^{32}P]$ATP in the assay from the $[\alpha^{32}P]$ATP standard c.p.m. *Provided* the assay volume is 100 μl, and the $[\alpha^{32}P]$ATP standard was prepared using a 1% dilution of the assay mix, the specific activity is given by dividing the ^{32}P c.p.m. by the assay [ATP] (in μM). The units of specific activity are c.p.m. per picomole.

3. Calculate an assay constant by multiplying the specific activity by the protein content (mg) of each assay tube. The units of the assay constant are c.p.m. per picomole·milligram.

4. Calculate the spillover of ^{32}P c.p.m. into the ^{3}H channel from the $[\alpha^{32}P]$ATP standard c.p.m. This is given by the ^{3}H c.p.m. divided by the ^{32}P c.p.m. from the $[\alpha^{32}P]$ATP standard.

5. For each sample, the $[^{32}P]$cAMP generated during the assay is calculated as follows: The units are picomoles cAMP per milligram membrane protein.

$$\frac{[\text{Sample } ^{32}\text{P c.p.m.}] \times [^3\text{H cAMP recovery standard c.p.m.}]}{\text{assay constant} \times [\text{Sample } ^3\text{H c.p.m.} - (\text{Sample } ^{32}\text{P c.p.m.} \times \text{spillover})]}$$

6. Adenylate cyclase activity is obtained by dividing this figure by the assay time.

---

i. Example

You have assayed adenylate cyclase activity for 20 min of a membrane preparation containing 30 μg protein per tube using 500 μM ATP in the assay. You obtain the following results:

Tube	^{3}H c.p.m.	^{32}P c.p.m.
Background	20	15
[^{3}H]cAMP recovery standard	21 020	17
[α^{32}P]ATP standard	321	8035
Sample	16 820	755

After background subtraction, these results become:

Tube	^{3}H c.p.m.	^{32}P c.p.m.
[^{3}H]cAMP recovery standard	21 000	(2)
[α^{32}P]ATP standard	301	8020
Sample	16 800	740

The specific activity of [α^{32}P]ATP in the assay is 8020 divided by 500

$$= 16.04 \quad \text{c.p.m. per picomole}$$

The assay constant is $16.04 \times 0.03 = 0.4812$ c.p.m. per picomole·milligram
The spillover is 301 divided by 8020 $= 0.0375$
The cAMP generated is therefore $[740 \times 21\,000]$ divided by $0.4812 \times [16800 - (740 \times 0.0375)]$

$$= 1925 \text{ pmol cAMP per milligram}$$

Adenylate cyclase activity is therefore 96.3 pmol cAMP per milligram per minute

This figure contains some component due to non-specific carryover of [α^{32}P]ATP, reflecting the failure of the column procedure to eliminate it completely from the samples. The zero time samples quantify this effect, and an apparent cyclase activity is calculated for the zero time samples and subtracted from each sample result. This correction should be very small, but may be important where activity is low or when the columns need regenerating with NaOH as described below.

It is a simple matter to program a microcomputer or even a programmable calculator to perform this calculation.

3.3.5 Maintenance of columns

Should the columns dry out when left for several days between experiments, it may be helpful to expel air from the ion-exchange materials by dipping the tip of the columns, in their racks, into a tray of distilled water. They will fill by capillary action. The preparation procedures detailed above can then be performed; the wet columns will not become air-locked.

Over a period of time the recovery of [^{3}H]cAMP in the assay may deteriorate. Optimal performance in the assay gives, for the elution volumes described above, a recovery of about 70%. At the same time the flow rate of the Dowex columns may also deteriorate. This results from denatured protein bound to the columns. The performance of the columns can be restored by washing with 10 ml of 1 M NaOH which will remove the protein, followed as soon as possible by 3 × 10 ml of water and then 10 ml of 1 M HCl. It is important to keep the time of exposure of the columns to basic conditions to a minimum since Dowex 50 is somewhat base labile. If relatively highly purified membrane preparations are used, the activity is such that only a few micrograms of protein are required for each determination. In this case, base washing is only required every 20–30 assays. If cruder preparations are used (for example, whole cell homogenates), and in particular if the preparation is of bacterial origin and still contains elements of the cell wall, much more frequent base treatment will be required. After five or six base treatments it is advisable to replace the Dowex in the columns, a rather tedious process but in most cases one which is required only once a year or so. If these procedures do not restore [³H]cAMP recovery, consider the possibility that the radiolabel is degraded; a significant proportion of the [³H]cAMP recovery standard c.p.m. may be [³H]AMP or ³H₂O in material more than a few months old.

The performance of the alumina columns will also eventually deteriorate. This becomes apparent by the zero time ³²P c.p.m. increasing. If they begin to rise the alumina should be replaced. It should, however, last for at least 100 determinations.

3.3.6 Safety considerations

The water bath in which the incubations are performed provides excellent shielding against β emissions which are lower than horizontal. We dispense the radioactive assay mix inside a box (about 70 × 40 × 40 cm) constructed of 5-mm Perspex from which the top and one side are missing, the open side abutting the water bath. This box contains spillage, provides some additional shielding and ensures that the operator is kept at arm's length from the water bath so that exposure to β emissions above the horizontal is minimised.

The assay tubes are held in solid Perspex blocks drilled to accommodate the tubes snugly; these provide effective shielding during handling procedures.

Waste eluate from the columns should ideally be run directly into a sink with running water to wash it down the drain. Most of the ³²P in the incubation will run to waste so the sink must be authorized for the appropriate amount of disposal. Stainless steel sinks should be avoided since the acid used to wash the Dowex columns will cause corrosion. Also radioactivity will tend to adsorb to the sink. Ideally, an epoxy resin sink should be used which minimizes adsorption of the radioactivity. In the UK these can be obtained from Simmons Mouldings Ltd., Parkside, Coventry CV1 2NE, who have a range of standard sinks or will fabricate a sink to specific requirements. A screen of 13-mm Perspex should be located in front of the columns to prevent exposure to radiation while the columns are running. This is not as severe a hazard as might be expected since the columns themselves provide effective shielding from ³²P radiation adsorbed to the column.

If all these precautions are followed, horizontal exposure to radioactivity should be undetectable with a standard laboratory monitor.

4. cAMP and adenylate cyclase by metabolic labelling of ATP

4.1 Introduction

An alternative approach to the measurement of cAMP in cells is to label their ATP pool by incubation with [^{3}H]adenine. This leads to formation of [^{3}H]cAMP which, in principle, might be separated from other radiolabelled materials and measured. In conjunction with ATP determinations, such measurements might yield absolute estimates of intracellular cAMP concentration which rival or exceed the sensitivity afforded by RIA, so that they might be undertaken on very small numbers of cells, or in cell systems which are not amenable to the approaches given in Sections 2 and 3 above. Detailed below is a development of the metabolic labelling method which although more complex than the above methods and requiring a greater investment in racks and columns, is reasonably economical of materials and time especially when large numbers of samples are to be determined.

4.2 Principle of the method

Cells are labelled with [^{3}H]adenine so that their ATP pool becomes radioactive. They are then treated with hormones or other agents which activate adenylate cyclase so that [^{3}H]cAMP is generated. [^{3}H]ATP and [^{3}H]cAMP are then separated from other radiolabelled materials (notably [^{3}H]adenine, but also adenosine, AMP, and ADP) and estimated by liquid scintillation. For cAMP, separation is achieved using a three-column chromatographic procedure, based on the separation used in *Protocol 11*, but with an additional Dowex-1 ion-exchange step to remove residual [^{3}H]adenine, which is in large excess and carries through the basic procedure.

The first column is of Dowex 50 ×4 200, as for *Protocol 11*, but has 2.5 ml bed volume. The second column is of alumina, and is as used in *Protocol 11*. The third column is of Dowex-1 in the chloride form, an anion-exchanger which binds [^{3}H]cAMP well so that [^{3}H]adenine can be washed off and the [^{3}H]cAMP recovered by elution with HCl. The recovery of [^{3}H]cAMP is quantified by including [^{14}C]cAMP in trace quantities at the outset.

The effluent from the first stage of the Dowex 50 column contains most of the [^{3}H]ATP. This is recovered, and applied to a fourth column which is again of Dowex-1 chloride. This separates ATP from contaminant ADP and other nucleotides by differential elution with HCl.

A complication of the method is that it relies on dual label counting of ^{3}H and ^{14}C, and that the different fractions generated from each sample will have somewhat different composition. It is essential, therefore, that an effective dual label d.p.m. option is available on the scintillation counter. We consider that some error is inevitably introduced by the approximations of the d.p.m. calculation. *Protocols 13* and *14* were developed for trypanosomes in suspension, and we anticipate that they could readily be applied to other cell suspensions. Some modification would be necessary to adapt the method to cells in monolayer culture, especially at the solubilization stage where large volumes of fluid might be generated. However, cells grown in microtitre plates might be assayed in this way with very little modification of the procedure.

Protocol 13. Metabolic labelling with [^{3}H] adenine

You will need:

- samples of cells in adenine-free medium
- 2-[^{3}H]adenine
- bench centrifuge
- stop solution (see *Protocol 10*)
- [^{14}C]cAMP in water (60 000 d.p.m. per millilitre)
- water baths, etc.

1. Incubate the cells at 37 °C with 100 µCi/ml [^{3}H]adenine
2. After a suitable time period, e.g. 1 h, centrifuge the cells and wash them free of [^{3}H]adenine by resuspending in ice-cold medium, and centrifuging again. Repeat the wash twice more.
3. Dispense out the cells in 50 µl medium. Incubate at 37 °C for suitable times with agents such as hormones, phosphodiesterase inhibitors, etc.
4. Add 50 µl stop and mix well.
5. Add 50 µl [^{14}C]cAMP (3000 d.p.m.) and then 850 µl water. Mix well.

The times and temperatures of incubations will need to be established by experiment, as will the concentration of cells. For trypanosomes, we have used 10^7 organisms/millilitre for both labelling and subsequent incubations. Similarly, it is possible that a lower concentration of [^{3}H]adenine could be used, which would improve the discrimination of the separation procedures, but this would also need to be determined by experiment. In other cells, it is possible that [^{3}H]adenosine might be a better source of radiolabel; our preliminary findings suggest that this might be true for hepatocytes. You would need to test this empirically to determine which gives greatest incorporation into the ATP pool, and which equilibrates fastest.

Protocol 14. Separation of metabolically labelled nucleotides

You will need:

- 4 racks of columns as described in *Figure 4*
- rack 1 is loaded with a 2.5-ml bed of Dowex 50
- rack 2 is loaded with 0.8 g of neutral alumina
- racks 3 and 4 each contain a 1-ml bed of Dowex 1, chloride form
- suitable racks of vials to collect eluates
- dispensers containing the following reagents: (a) water, (b) 0.3 M imidazole, pH 7.4, (c) 3.16 mM HCl, (d) 40 mM HCl, (e) 1 M HCl

A. *Method for [³H]cAMP separation*

1. Locate rack 1 (Dowex 50) on rack 4 (Dowex 1).
2. Load the stopped, diluted sample (*Protocol 12*) and wash with 5 ml water. This loads [³H]cAMP into Dowex 50, and washes [³H]ATP into Dowex 1.
3. Locate rack 1 on rack 2 (alumina), wash the sample into the alumina with 10 ml water. This loads [³H]cAMP into the alumina.
4. Remove and regenerate rack 1.
5. Add 0.5 ml 3 M imidazole to rack 2, and allow to run to waste.
6. Locate rack 2 on rack 3 (Dowex 1), and wash the sample from the alumina with 2.5 ml imidazole. This loads [³H]cAMP into Dowex 1.
7. Remove rack 2 (alumina).
8. Wash rack 3 with 1 ml water, then 15 ml 3.16 mM HCl. This washes [³H]adenine to waste.
9. Locate rack 3 on scintillation vials and elute [³H]cAMP with 1.5 ml 1 M HCl.
10. Add scintillant to the vials, mix well and count.

B. *Method for [³H]ATP separation*

1. Rack 4 (Dowex 1) contains the effluent from rack 1 (Dowex 50)
2. Wash with 30 ml 40 mM HCl.
3. Locate rack 4 on scintillation vials.
4. Elute [³H]ATP with 1.5 ml 1 M HCl.
5. Add scintillant to the vials, mix well and count.

The procedure should be calibrated with appropriate standard materials to ensure that the volumes given here effectively discriminate between [³H]adenine, [³H]cAMP, and [³H]ATP. For the calibration, each of the effluents should be collected and sampled before moving on to the text stage. Thereafter, a procedure similar to the above with adjusted volumes should suffice.

Regeneration of the Dowex 50 and alumina columns is as described in *Protocol 11*, i.e. they should be washed with 1 M HCl and imidazole respectively. For Dowex 1 columns, a base wash needs to be performed first, as in Section 3.3.5, followed by water then HCl, and finally a water wash.

4.2.1 Calculation

The yield of $[^{14}C]$cAMP is calculated for each sample. The yield of $[^{3}H]$cAMP generated in each sample is then corrected for losses during separation.

The method assumes quantitative yield of $[^{3}H]$ATP; we routinely find better than 95% recovery results from this procedure. The purpose of separating $[^{3}H]$ATP at all is to relate the $[^{3}H]$cAMP c.p.m. to the $[^{3}H]$ATP c.p.m., which may be important if it is considered that the specific activity of $[^{3}H]$ATP might change during the experiment, particularly in response to the treatments applied to the cells. Formal determination of the ATP content of the cells, using standard biochemical methods such as the luciferase assay (23), would enable the specific activity of ATP to be determined, and hence an absolute estimate of cAMP levels to be produced.

5. Guanylate cyclase and cGMP

There is substantial and obvious structural analogy between the substrates and products of adenylate and guanylate cyclases, although the structures of the enzymes themselves appear to be very different. A role for G-protein in the regulation of guanylate cyclase has not yet been determined, and indeed, seems unlikely. However, the techniques for determining their activity using radiolabelled substrates are substantially similar to those described above. A soluble form of guanylate cyclase can be activated in smooth muscle and other cells by nitric oxide (see ref. 24) and a resurgence of interest in guanylate cyclase has followed the realization that platelet activation can be inhibited by endothelium-derived relaxation factor, now known to be nitric oxide (25). For these reasons we consider it worth mentioning that an assay for guanylate cyclase can be performed using the same equipment as we have detailed above for adenylate cyclase determination. It has been stated (26) that Dowex 50 and alumina columns identical to those used in *Protocol 11* will satisfactorily separate cGMP from GTP, so that the only modification of *Protocols 10* and *11* needed is to substitute $[\alpha^{32}P]$GTP, GTP and cGMP recovery standards for their adenine nucleotide counterparts. Although our experience of this area is limited, we consider that cGMP binds less well to Dowex 50 than cAMP, and we therefore prefer to use a larger resin bed, for example 2.5 ml, as detailed in *Protocol 13*. Elution profiles would need to be checked for the specific column geometries in use.

Analogous RIA methodology (26) for the determination of cGMP is also routinely available from various radiochemical suppliers, and again, since they supply detailed protocols for this assay we have not provided details of these.

References

1. Birnbaumer, L. Abramowitz, J., and Brown, A. M. (1990). *Biochim. Biophys. Acta*, **1031**, 163–224.
2. Lefkowitz, R. J. and Caron, M. G. (1988). *J. Biol. Chem.*, **263**, 4993–6.
3. Premont, R. T. and Iyengar, R. (1990). In *G proteins* (ed. L. Birnbaumer and R. Iyengar), pp. 148–78. Academic Press, San Diego.
4. Rodbell, M., Lin, M. C., Salomon, Y., Londos, C. D., Harwood, J. P., Martin, B. R., Rendell, M., and Berman, M. (1975). *Adv. Cyclic Nuc. Res.*, **5**, 3–29.
5. Heideman, W. and Bourne H. R. (1990). In *G proteins* (ed. L. Birnbaumer and R. Iyengar), pp. 17–42. Academic Press, San Diego.
6. Neer, E. J. and Clapham, D. E. (1990). In *G proteins* (ed. L. Birnbaumer and R. Iyengar), pp. 42–63. Academic Press, San Diego.
7. Krupinski, J., Coussen, F., Bakalyar, H. A., Tang, W.-E., Feinstein, P. G., Orth, K., Slaughter, C., Reed, R. R., and Gilman, A. G. (1989). *Science*, **244**, 1558–64.
8. Wong, S.K.-F. and Martin, B. R. (1985). *Biochem. J.*, **231**, 39–46.
9. Brown, B. L., Albano, J. D. M., Ekins, R. P., Sgherzi, A. M., and Tampion, W. (1971). *Biochem. J.*, **121**, 561–2.
10. Palmer, S. and Wakelam, M. J. O. (1990). In *Methods in inositide research* (ed. R. F. Irvine), pp. 127–34. Raven Press, New York.
11. Santa-Coloma, T. A., Bley, M. A., and Charreau, E. H. (1987). *Biochem. J.*, **245**, 923–4.
12. Houslay, M. D., Metcalfe, J. C., Warren, G. B., Hesketh, T. R., and Smith, G. A. (1976). *Biochim. Biophys. Acta*, **436**, 489–94.
13. Knowles, B. H. and Farndale, R. W. (1988). *Biochem. J.*, **253**, 235–41.
14. O'Neill, L. A. J., Stylianou, E., Edbrooke, M. R., Farndale, R. W., and Saklatvala, J. (1991). In *Methodological surveys in biochemistry and analysis*, Vol. 21, (ed. E. Reid, G. M. W. Cook, and J. P. Luzio), pp. 31–8. Royal Society of Chemistry, Cambridge.
15. Farndale, R. W. and Murray, J. C. (1986). *Biochim. Biophys. Acta*, **881**, 46–53.
16. Samuelsson, B., Grantstrøm, E., Green, K., Hamberg, M., and Mammarstrøm, S. (1975). *Annu. Rev. Biochem.*, **44**, 669–95.
17. Farndale, R. W., Sandy, J. R., Atkinson, S. J., Pennington, S. R., Meghji, S., and Meikle, M. C. (1988). *Biochem. J.*, **252**, 263–8.
18. Frandsen, E. K. and Krishna, G. (1976). *Life Science*, **18**, 529–42.
19. Salomon, Y., Londos, C., and Rodbell, M. (1974). *Anal. Biochem.*, **58**, 541–8.
20. Voorheis, H. P. and Martin, B. R. (1982). *Eur. J. Biochem.*, **123**, 371–6.
21. Stein, J. M., Kornberg, H. L., and Martin, B. R. (1985). *FEBS Lett.*, **182**, 429–34.
22. Farndale, R. W., Wong, S.K.-F., and Martin, B. R. (1987). *Biochem. J.*, **242**, 637–43.

23. Lundin, A., Rickardsson, A., and Thore, A. (1976). *Anal. Biochem.*, **75**, 611–20.
24. Murad, F. (1988). In *Colloque INSERM*, Vol 165 (ed. J. Nunez, J. E. Dumont, and E. Carafoli), pp. 3–12. Colloque INSERM/John Libby Eurotext, London.
25. Radomski, M. W., Palmer, R. M. J., and Moncada, S. (1990). *Proc. Natl. Acad. Sci. USA*, **87**, 5193–7.
26. Mittal, C. K. (1986). *Methods Enzymol.*, **132**, 422–34.

5

Inositol lipids and phosphates

PHILIP P. GODFREY

1. Introduction

There is now much evidence that stimulation of inositol phospholipid metabolism via a phosphoinositide-specific phospholipase C (PIC) is the signal transduction pathway for a wide variety of receptors in eukaryotic cells. Receptors are thought to be coupled to PIC through a guanine nucleotide binding protein termed G_p, by a mechanism analogous to coupling of receptors with adenylate cyclase. Recent work suggests this G-protein may be the same as G_q earlier identified by molecular biology techniques. The initial receptor-coupled event is a PIC-mediated breakdown of phosphatidylinositol 4,5-bisphosphate (PIP_2) to give the two second messengers inositol 1,4,5-trisphosphate ($1,4,5-IP_3$) which mobilizes intracellular calcium and 1,2-diacylglycerol (DAG) which activates protein kinase C. $1,4,5-IP_3$ is then either phosphorylated to inositol 1,3,4,5-tetrakisphosphate ($1,3,4,5-IP_4$) or dephosphorylated to $1,4-IP_2$, the reactions catalysed by a 3-kinase or 5-phosphatase respectively. $1,3,4,5-IP_4$ is subsequently dephosphorylated to $1,3,4-IP_3$ by the same 5-phosphatase that degrades $1,4,5-IP_3$ and is then further metabolized to a variety of inositol bis- and monophosphates; one key area of research in recent years has been to try and elucidate these complex metabolic interactions. One feature of particular note in the catabolic pathways for $1,4,5-IP_3$ is that two key enzymes, inositol polyphosphate 1-phosphatase and inositol monophosphatase, are specifically inhibited by lithium ions in the millimolar range. Stimulation of cells in the presence of lithium results in a substantial build-up, particularly of inositol monophosphates, and this has been used to provide the basis for a simple assay of phosphoinositide turnover (ref. 1; see later). It has also been proposed that this effect of lithium is the biochemical event underlying the action of lithium in the therapy of manic-depressive illness.

In this chapter we will give protocols for the various techniques we have used to investigate inositol lipid and phosphate metabolism, including techniques of lipid analyses, and both radioactive and mass assays of inositol phosphates. There are several other techniques which we have not used ourselves, that readers may find useful, and I will provide the appropriate references for these methods. Hopefully this should cover the vast majority of techniques that anyone would want to use. Several other reviews of methods for analysis of inositides have recently appeared (2–4) and these are recommended.

The type of analysis procedure used will depend very much on the type of question that the investigator wishes to ask. For example if the experiment is to correlate changes in 1,4,5-IP$_3$ or 1,3,4,5-IP$_4$ levels with increases in cytoplasmic calcium then analysis of mass amounts of these metabolites would be most appropriate. If effects of receptor stimulation on metabolic interconversions of inositol phosphates are being studied then separation using dowex column chromatography or HPLC will be necessary. When a simple measure of the extent of receptor activation is required (for example, when investigating the pharmacology of a particular receptor coupled to PI hydrolysis), the assay of a total inositol phosphate (IP) fraction in the presence of LiCl provides a more accurate estimation of the extent of receptor activation than does measurement of a single metabolite.

2. Incubation buffer

Our standard buffer is a Krebs–Ringer–Hepes buffer [composition (mM): NaCl 130, KCl 5, MgSO$_4$ 1.2, CaCl$_2$, 1.2, Hepes 20, Na$_2$HPO$_4$ 1.2, and glucose 10; pH 7.4] equilibrated with 100% O$_2$ at 37 °C, though the use of this is by no means *de rigeur*. When 10 mM LiCl is present this replaces the equivalent amount of NaCl.

Most experiments involve the labelling of the cell preparation with either [^{32}Pi] or [^{3}H]inositol. The precise amounts of label required will depend very much on the cells being used. Generally, with [^{3}H]inositol we have tended to use 2 μCi/ml for experiments measuring a total inositol phosphate (IP) fraction and anything from 20 μCi/ml up when measuring individual IP isomers. Of course, when assaying mass levels of inositol phosphates, no radioactivity is required.

With tissue slices we normally pre-incubate the cells for an hour in buffer before adding the label in a small volume of buffer (without LiCl). This is then usually incubated in a shaking water bath at 37 °C; the tissue should be regularly oxygenated. This protocol does not label cells to equilibrium, which requires 24–48 h; however, it is exceedingly difficult to maintain viability of tissue slices for that length of time. Generally, cultured cells are used when prolonged labelling periods are necessary. Cells can then be labelled with [^{3}H]inositol for whatever time is required (usually 24 h); inositol-free medium is normally used since this allows a higher specific activity of label to be attained.

3. Extraction of inositol lipids and phosphates from tissues

The commonest method to quantitatively extract inositol lipids from tissues uses acidified chloroform/methanol (5). For extraction of the water-soluble inositol phosphates several methods are regularly used, including chloroform/ methanol, chloroform/methanol/HCl (as for the lipids), perchloric acid (PCA) followed by neutralization by KOH or by FREON/octylamine, and trichloroacetic acid (TCA) with neutralization by diethyl ether. All of these methods are simple and consistent. For extraction of acid-labile cyclic inositol phosphates a mixture of phenol, chloroform and methanol can be used (6). No extraction procedure will be totally satisfactory in all respects and the procedure of choice will depend on what measurements are going to be required. It is always advisable to check efficacy of extraction by following recoveries of radioactive standards added to tissue extracts following termination of the incubation.

3.1 Extraction of lipids

The procedure for extracting lipids (see *Protocol 1*) is slightly different to that for extraction of inositol phosphates (*Protocol 2*) requiring a washing step to remove contaminating water-soluble radioligand. This procedure provides a consistent and reproducible extraction of all the inositol lipids. Neutral chloroform/methanol will extract PI, though it will not provide quantitative extraction of the polyphosphoinositides.

Protocol 1. Extraction of phospholipids

Materials
- chloroform/methanol (1:2, v/v)
- 6 M HCl
- chloroform
- methanol
- 2 M KCl
- methanol/1 M HCl (1:1, v/v)

Method
1. Terminate the incubation at the required time with 3 vol. of chloroform/ methanol, containing 20 μl of 6 M HCl. Allow to extract for 30 min, then add 1 vol. of chloroform and 1 vol. of 2 M KCl. The initial extract should be a single phase, which should then split into two after the addition of

Protocol 1. *Continued*

chloroform and KCl. Although the precise volumes used are not important the ratio of chloroform/methanol/KCl should be kept constant.

2. Vortex and centrifuge (1500 *g* for 10 min) the extract and remove the lower phase.

3. Wash the upper phase with 2 vol. of chloroform by vortexing then centrifuge again.

4. Combine the lower phases and wash them with 1 ml of methanol/HCl, then dry down under nitrogen.

3.2 Chloroform/methanol extraction and analysis of a total inositol phosphate fraction

This procedure (*Protocol 2*) is based on that originally described in ref. 2. Chloroform/methanol will not quantitatively extract the highly phosphorylated IPs (IP_3 and above), though as these generally make up < 5% of a total IP fraction this does not significantly affect results. Acidification of the chloroform/methanol with HCl at the initial extraction step will provide a more reproducible extraction of the higher IPs though samples will then need to be neutralized prior to anion–exchange chromatography.

Protocol 2. Extraction with chloroform/methanol and analysis of a total water soluble inositol phosphate fraction

Materials

- chloroform/methanol (1:2)
- chloroform
- 15-ml conical test-tubes
- Dowex AG1X8 anion-exchange resin (200–400 mesh, formate form)
- 1 M ammonium formate/0.1 M formic acid

Method

1. Stop incubations (310 μl) with 0.94 ml (3 vol.) chloroform/methanol and then split the phases with 0.31 ml water and 0.3 ml chloroform. Vortex and centrifuge (1500 *g* for 10 min), then add 0.75 ml of upper phase to 2.25 ml water in a conical test-tube.

2. Add 0.5 ml of packed Dowex AG1X8 and then vortex. Allow gel to settle. Pour off supernatant, then add 3 ml water. Vortex and allow to settle; repeat this operation three times.

3. Extract the inositol phosphates with 0.5 ml of ammonium formate/formic acid, then vortex and allow to settle. Remove 0.4 ml of the supernatant into a 20-ml scintillation vial. Then repeat this, taking off 0.5 ml and combine this fraction with the other extract.

4. Add 10 ml of scintillant to each combined extract and count for radioactivity.

3.3 Extraction of inositol phosphates with perchloric acid

Two distinct methods have been developed which employ perchloric acid (PCA) as an extracting agent (*Protocol 3*). It has been suggested that to ensure good recoveries of radiolabelled inositol phosphates a small amount of phytic acid hydrolysate (7) should be added with the PCA.

Protocol 3. Extraction of inositol phosphates with perchloric acid

Materials

- 4.5% perchloric acid (PCA, ice-cold)
- 0.5 M KOH/9 mM sodium tetraborate
- 10% PCA (ice-cold)
- 1,1,2-trichlorotrifluoro-ethane (FREON)/tri-*N*-octylamine (1:1)

Method

1. Incubate cells as normal, then to stop the reaction add 0.6 ml of 4.5% PCA to 0.3 ml of cell preparation. Leave for 10 min on ice, then vortex and centrifuge (2000 g for 5 min). Remove 0.7 ml and neutralize this with a sufficient volume of KOH/sodium tetraborate to bring the pH to 8–9. Centrifuge to remove precipitated potassium perchlorate and use the supernatant for inositol phosphate analysis.

2. Stop the incubation (0.3 ml) with 0.3 ml of ice-cold 10% PCA, add 20 μl of 1% EDTA and leave on ice for 10 min. Vortex and centrifuge. Transfer 0.5 ml of supernatant to an Eppendorf tube and then add 0.6 ml of FREON/octylamine mixture (made up freshly). Vortex vigorously for at least 15 sec, then centrifuge. The mixture should resolve into three phases. The upper phase, which contains the inositol phosphates, can then be analysed.

If the extracts are being used for mass analysis of inositol phosphates (*Protocol 9* and *10*) ensure that the assay buffer used for the standards and blanks has been through the extraction procedure.

3.4 Extraction with trichloroacetic acid

This method (*Protocol 4*) is similar to those with PCA in terms of ease, reproducibility, and simplicity. It is worth ensuring that all the TCA has been extracted into the ether phase; simply check the pH has risen above 5.

Protocol 4. Extraction of inositol phosphates with trichloroacetic acid

Materials

- 30% trichloroacetic acid (TCA, ice-cold)
- water-saturated diethyl ether
- bath of methanol/dry-ice
- 10 mM NaHCO$_3$/5 mM EDTA

Method

1. Stop the incubations (250 μl) with 50 μl TCA, transfer the samples to Eppendorf tubes, then vortex and centrifuge (2000 *g* for 5 min).
2. Remove 250 μl of supernatant into a glass test-tube and add 1.0 ml of ether. Then place in the methanol/dry-ice bath. This freezes the aqueous sample whilst leaving the ether liquid; this can be poured off.
3. Repeat the ether extraction a further three times. Refreeze the samples and dry down under vacuum. Dried extracts can be resuspended in NaHCO$_3$/EDTA or other buffer as required.

---

4. Separation and analysis of phosphoinositides

Although there are a wide variety of techniques available for separation of the lipid intermediates of the phosphoinositide cycle, including paper or column chromatography (see *Protocols 2 and 3*), thin-layer chromatography (TLC), and deacylation, followed by separation of the water-soluble products have proved to be the methods of choice for the majority of determinations. These latter techniques are detailed below. I will also describe a simple method I have developed for the measurement of CDP-diacylglycerol (CDP-DG) which is an intermediate in the lipid PI cycle and whose concentration can change substantially following stimulation, especially in the presence of lithium.

4.1 Thin-layer chromatography (TLC)

We normally use TLC for the separation of [^{32}P]labelled lipids from [^{32}P]PI-labelled cells, since this provides a convenient way to analyse all phospholipid species in a single sample. We have never adequately separated all phospholipids, including polyphosphoinositides, on a single TLC system and so samples are split into two, with one plate for polyphosphoinositides and one for the other lipids. A wide variety of TLC solvent systems have been used over the years (2); those listed below have proved most worthwhile. 0.25-mm-thick E. Merck silica gel 60 plates that have been oxalate (1% potassium oxalate) impregnated (to cut down on streaking) are generally used. An aliquot of dried lipid is resuspended in chloroform/methanol (9:1) and spotted on the plate approximately 1 cm from the bottom.

Solvent systems used:

(a) for polyphosphoinositides:

a one-dimensional system of chloroform:methanol:88% ammonia:water (90:90:7:20, by volume)

(b) for other phospholipids

i. a one-dimensional system of chloroform:methanol:acetone:acetic acid:water (40:15:13:12:8, by volume).

ii. a double solvent system, developed in the same direction, of chloroform:methanol:88% ammonia (400:10:1) followed by chloroform:methanol:88% ammonia:water (65:35:2:3).

iii. a double solvent system, developed in the same direction, of chloroform:methanol:acetic acid:water (first 40:10:10:1, then 120:4:6:19:3).

The R_f values differ for each solvent system and non-radioactive standards should therefore be run to verify the position of each compound. The lipids are normally visualized using the Sigma Chemical Company molybdenum spray reagent or iodine vapour. The positions of the radioactive spots can be easily determined by autoradiography and the appropriate lipids can then be scraped and counted for radioactivity.

Lipid samples can also be taken for estimation of PI. We scrape the lipids into borosilicate test-tubes and digest for 2 h in 0.2 ml 70% (v/v) perchloric acid at 200 °C. The samples are then made up to 2 ml and centrifuged to remove the silica. The supernatant can then be removed and assayed for PI (8).

4.2 Deacylation of lipids and separation of water-soluble glycerophosphoinositols

For separation of [³H]inositol labelled inositol lipids (*Protocol 5*) we have used the method described by Creba *et al.* (9). The water soluble derivatives of all three inositol lipids can easily be quantified.

Protocol 5. Separation of water-soluble deacylated lipids

Materials

- chloroform
- methanol
- 1 M NaOH in methanol/water (19:1, v/v)
- 0.25 M boric acid
- 0.18 M ammonium formate/5 mM sodium tetraborate
- 0.4 M ammonium formate/1.0 M formic acid
- 1.0 M ammonium formate/0.1 M formic acid
- 2.0 M ammonium formate/0.1 M formic acid
- 1 ml column of Dowex AG1X8 (200–400 mesh) anion-exchange resin.

Method

1. Dissolve dried lipids in 1 ml chloroform then add 0.2 ml methanol and 0.2 ml 1 M NaOH in methanol:water. Digest at room temperature for 20 min then add 0.6 ml methanol and 0.6 ml water. Vortex and centrifuge (1500 g for 10 min).

2. Remove 1 ml of upper phase and neutralize with boric acid (0.25 M). Then dilute samples to 5 ml with sufficient ammonium formate/sodium tetraborate to give final concentrations of 0.18 M and 5 mM respectively.

3. Load the mixture on to a 1 ml Dowex AG1X8 column and elute with a further 15 ml of 0.18 M ammonium formate/5 mM sodium tetraborate. Combine the eluate with that from the original column loading and take a sample for scintillation counting. This contains labelled glycerophosphoinositol, the deacylated product of PI.

4. Glycerophosphoinositol phosphate (from PIP) is then eluted with 20 ml of 0.4 M ammonium formate/0.1 M formic acid and glycerophosphoinositol bisphosphate (from PIP$_2$) with 20 ml of 1 M ammonium formate/0.1 M formic acid. Samples from each fraction can then be added to scintillation fluid and counted.

5. Columns are regenerated with 2 M ammonium formate/0.1 M formic acid and then washed with water. They can be used for 4–5 separate runs.

Inositol lipids phosphorylated at the 3-position have recently been identified and the deacylated products of these lipids can be separated from the conventional 4-phosphorylated lipids by HPLC.

4.3 Assay of CDP-diacylglycerol (CDP-DG)

When cells are stimulated in the presence of lithium there is a reduction in intracellular inositol levels and this is reflected in a reduction in PI resynthesis and a compensatory accumulation of the other precursor of PI, namely CDP-DG. We have developed a simple assay for CDP-DG involving pre-labelling of endogenous CTP with cytidine, which then accumulates as CDP-DG upon stimulation in the presence of LiCl (*Protocol 6*). The rise in CDP-DG can be reversed by pre-incubation with myo-inositol, and can provide a useful index of cellular inositol concentrations (10). Since CDP-DG is the only lipid that labels with cytidine the samples can be analysed without the need for TLC.

Protocol 6. Assay of CDP-diacylglycerol

Materials

- Krebs–Hepes buffer (see Section 2) wth 10 mM LiCl
- chloroform:methanol (1:2, vol./vol.), chloroform
- methanol/1 M HCl (1:1 vol./vol.)
- [^{3}H]cytidine

Method

1. Incubate 50 µl of cells/tissue slices in 250 µl of buffer containing 2 µCi/ml labelled cytidine and 10 mM LiCl, with shaking, for 15 min at 37 °C. Agonist (10 µl) is then added; antagonists should be added 5 min prior to agonist addition.

2. Gas samples with O_2 then cap tubes and incubate for a further 60 min (or other times as required). Stop the reaction with 0.94 ml chloroform/methanol (1:2), then split the phases with 0.31 ml chloroform and 0.31 ml water.

3. Vortex thoroughly then centrifuge (1500 *g*, 10 min). Remove a 0.45 ml aliquot of the lower phase into a glass test-tube. Wash this with 1 ml of methanol/HCl, vortex, and centrifuge again. Then remove the bottom phase into a scintillation mini-vial and dry down.

4. Add 5 ml of scintillation fluid, vortex, then count for radioactivity.

---

5. Methods for analysis of inositol phosphates

There are a wide variety of methods for estimation of mass levels of inositol phosphates, of varying complexity (see refs 2–4). The commonest method involves analysis of mass levels of 1,4,5-IP$_3$ or 1,3,4,5-IP$_4$ by the use of a specific and selective binding protein (*Protocols 9* and *10*), and this will be described below. Radiolabelled inositol phosphates are usually separated by anion-exchange chromatography (*Protocol 7*) or HPLC; paper chromatography can also be used (see ref. 3) though this has now largely been superseded by the ion-exchange techniques. Gas chromatography and GC-mass spectography can also be used to measure mass levels of inositol phosphates (11) and sensitivities of under 1 pmol can be obtained. However, this method requires complex and expensive equipment and will not be described in detail here.

Some techniques can be applied to all inositol phosphates (this usually involves the use of radiolabelled precursors) whereas some will measure only a few. The method of choice depends on a number of factors including: (a), the inositol phosphate of interest; (b), the number of samples that need to be analysed; (c), the availability of material for analysis; (d), inital outlay for equipment and running costs.

5.1 Separation of radiolabelled inositol phosphates

The analysis of a total IP fraction (see Section 3.2) is simple, convenient, cheap and a large number of samples can be assayed; however, it tells you nothing about what inositol phosphates are produced following receptor activation. To resolve these phosphate isomers either anion-exchange chromatography with a gradient elution or HPLC is employed. The former method is simpler but does not provide the resolution obtainable with HPLC.

Protocol 7. Separation of inositol phosphates by anion-exchange chromatography

Materials

- Dowex AG1X8 anion-exchange resin (200–400 mesh, formate form)
- 60 mM ammonium formate/5 mM sodium tetraborate
- 0.2 M ammonium formate/0.1 M formic acid
- 0.4 M ammonium formate/0.1 M formic acid
- 0.8 M ammonium formate/0.1 M formic acid
- 1.2 M ammonium formate/0.1 M formic acid
- 2.0 M ammonium formate/0.1 M formic acid
- 1 ml column of Dowex AG1X8 (200–400 mesh)

Method

1. After the incubation is completed terminate the reaction with one of the acid extraction methods described in Section 3. The aqueous extract containing the inositol phosphates is then diluted to 10 ml with water.

2. Put this on a 1-ml Dowex AG1X8 column and elute. Then elute free inositol with a further 20 ml of water. Elute glycerophosphoinositol with 16 ml of 60 mM ammonium formate/5 mM sodium tetraborate. Inositol monophosphates are then eluted with 16 ml 0.2 M ammonium formate/0.1 M formic acid, inositol bisphosphates with 16 ml of 0.4 M ammonium formate/0.1 M formic acid, inositol trisphosphates with 8 ml of 0.8 M ammonium formate/0.1 M formic acid and inositol tetrakisphosphates with 8 ml of 1.2 M ammonium formate/0.1 M formic acid. A sample of each fraction can then be taken for scintillation counting.

3. Wash the column with 10 ml of 2.0 M ammonium formate/0.1 M formic acid followed by 20 ml water. They can then be used for 4–5 separate runs.

Check the elution profiles using radioactive standards. When higher salt concentrations are used (i.e. for elution of IP_3 and IP_4) the scintillation fluid will not always form a single phase with the sample; under these circumstances either add a little methanol to the mixture or predilute the eluate with water.

A simple enzymic method for separation of $[^3H]$-1,4,5-IP_3 and $[^3H]$-1,3,4-IP_3 has recently been developed (12), based on the observation that the 1,4,5-IP_3 5-phosphatase is a magnesium-dependent enzyme, whereas the 1,3,4-IP_3 4-phosphatase is Mg-independent. Thus, in the presence of EDTA 1,3,4-IP_3 is selectively degraded and the remaining radioactivity can then be attributed to 1,4,5-IP_3. The only drawback of this technique is that in cells that have been labelled for an extended period of time (> 24 h) other inositol trisphosphate isomers may contain significant amounts of radioactivity, though in general this does not appear to be important.

Protocol 8. Separation of $[^3H]$-1,4,5-IP_3 and $[^3H]$-1,3,4-IP_3

Materials

- a rat brain
- polytron
- 250 mM Hepes/2 mM $MgCl_2$ pH 7.4
- 250 mM Hepes/5 mM EDTA
- 10% perchloric acid
- FREON/octylamine (see *Protocol 3*)
- tissue sample extracted with FREON/octylamine

Protocol 8. *Continued*

Method

1. A crude preparation of rat brain cytosol is used as an enzyme source; a rat brain is homogenized (2 × 15 sec in a polytron, setting 6) in 250 mM Hepes/2 mM $MgCl_2$ pH 7.4, at a concentration of 25% (w/v) and then centrifuged at 4 °C for 90 min at 100 000 g. The supernatant is used as enzyme source.

2. Add 300 μl of top layer from a FREON/octylamine tissue extraction (see *Protocol 3*) to 50 μl 250 mM Hepes/5 mM EDTA and 17.5 μl of brain supernatant and continue the incubation at 37 °C for 30–60 min.

3. Stop the reaction with 370 μl of 10% PCA and then re-extract the samples with FREON/octylamine (*Protocol 3*). 1,4,5-IP_3 can then be separated using conventional dowex column chromatography (*Protocol 7*). 1,3,4-IP_3 content can be assumed to be the remaining counts in a total IP_3 fraction after subtraction of the 1,4,5-IP_3 radioactivity.

If individual inositol phosphate isomers need to be analysed (for example, when the metabolic interconversions of inositol phosphates are being studied), then HPLC should be the method of choice. This is the only method with the power to resolve all the different phosphates. Its disadvantages are that it is very time-consuming and expensive. Things have become a little easier recently with the advent of automatic sampling and on-line radioactivity detectors, though this has not made the process any quicker. However, its resolving power is unquestionable. We have never had access to the necessary HPLC equipment, so I have no personal experience of this technique. However, Dean and Beaven (3) have described their HPLC methodology in some detail and this should provide sufficient information for the beginner. Similarly, Batty *et al.* (13) have developed an HPLC method for separation of inositol phosphates in brain tissue slices.

5.2 Mass analysis of inositol phosphates

In the past 2–3 years specific binding assays for the measurement of both 1,4,5-IP_3 (14, 15) and 1,3,4,5-IP_4 (16, 17) mass have been developed, based on displacement of [^3H]-1,4,5-IP_3 or [^{32}P]-1,3,4,5-IP_4 from specific binding proteins in bovine adrenal cortex or rat cerebellum respectively (*Protocols 9* and *10*). The assays are very easy to do and large numbers can be processed simultaneously. The only drawbacks are the costs of the radiolabelled inositol phosphates and the time it takes to prepare the binding proteins. Time can be saved by preparing the latter in bulk and freezing down aliquots of membranes; we have found the adrenal protein to be stable for at least 6 months at −70 °C.

Protocol 9. Mass measurement of inositol 1,4,5-trisphosphate using a binding protein

Materials

- bovine adrenal glands
- 20 mM NaHCO$_3$
- incubation buffer (20 mM NaCl, 100 mM KCl, 1 mM EDTA, 1 mg/ml BSA, 20 mM Tris–HCl, pH 8.3)
- [^{3}H]inositol, 1,4,5 trisphosphate ([^{3}H] IP$_3$)
- inositol 1,4,5 trisphosphate (IP$_3$)
- polytron
- Eppendorf tubes
- microfuge

Method

I. Preparation of an IP$_3$ binding protein.

1. The adrenal cortex is dissected from bovine adrenal glands on ice.
2. The adrenal cortex is homogenized in 20 mM NaHCO$_3$ (20 vol.) using a polytron (3 × 15 sec bursts at setting 6).
3. The homogenate is centrifuged (1000 g, 10 min, 4 °C).
4. The supernatant is removed and stored and steps (b) and (c) repeated on the pellet.
5. The combined supernatants from step (d) are combined and centrifuged (38 000 g, 20 min, 4 °C).
6. The membrane pellet is resuspended in incubation buffer at 5–10 mg/ml.
7. Aliquots of the binding protein preparation are stored at −80 °C.

II. IP$_3$ mass determination.

1. The binding assays are performed on ice in a final volume of 250 μl.
2. For each assay add 25 μl [^{3}H] IP$_3$ (0.25–0.50 nM final).
3. Add differing concentrations of cold IP$_3$ (for the standard curve) or unknown sample (prepared as in *Protocol 3* or *4*).
4. To define non-specific binding perform assays which contain 10 μM cold IP$_3$.
5. Add 150 μl of binding protein to each assay. Vortex samples and incubate (10 min, 4 °C).

Protocol 9. *Continued*

6. Terminate incubations by centrifugation in a microcentrifuge (13 000 r.p.m. 10 min).

7. Discard supernatant.

8. Wash pellet twice with 1 ml water.

9. Treat pellet overnight with 100 μl tissue solubilizer.

10. Add 1 ml liquid scintillant and determine radioactivity.

11. Compare standard displacement curve with unknown and interpolate to obtain results.

Protocol 10. Mass measurement of inositol 1,3,4,5-tetrakisphosphate using a binding protein

Materials

• rat cerebellum
• 20 mM $NaHCO_3$/1 mM dithiothreitol
• [^{32}P] inositol 1,3,4,5-tetrakisphosphate
• inositol 1,3,4,5-tetrakisphosphate (1,3,4,5 IP$_4$)
• Whatman GF/B filters

Method

I. Preparation of an 1,3,4,5 IP$_4$ binding protein.

1. Cerebella are removed from capitated rats and homogenized in ice-cold 20 mM $NaHCO_3$/1 mM dithiothreitol (20 vol.).

2. The homogenate is centrifuged (38 000 g, 20 min, 4 °C).

3. The supernatant is discarded, the pellet resuspended in 20 vol. 20 mM $NaHCO_3$/1 mM dithiothreitol and step (b) repeated.

4. The supernatant is discarded and the pellet resuspended as above at 5–10 mg/ml.

5. Aliquots of the crude binding protein preparation are stored at −80 °C.

II. 1,3,4,5-IP$_4$ mass determination.

1. Assays are performed in a final volume of 160 μl on ice.

2. Each assay contains 25 mM sodium acetate, 25-mM K_2HPO_4, 2 mM EDTA pH 5.0.

3. Add approx. 15 000 d.p.m. [^{32}P]1,3,4,5 IP$_4$ and either differing concentrations of cold 1,3,4,5-IP$_4$ or unknown fractions (prepared as in *Protocols 3* and *4*).

4. To define non-specific binding perform assays which contain 10 μM cold 1,3,4,5-IP$_4$.

5. Add binding protein to initiate reaction and incubate on ice for 30 min.

6. Occassionally vortex samples during the incubation period.

7. Terminate incubations by rapid filtration through Whatman GF/B filters.

8. Remove free [^{32}P]1,3,4,5-IP$_4$ by washing the filter three times with 3 ml 25 mM sodium acetate, 25 mM K$_2$HPO$_4$, 2 mM EDTA pH 5.0.

9. Determine radioactivity remaining on the filter by liquid scintillation counting.

10. Compare standard displacement curve with unknown and interpolate to obtain results.

Inositol 1,4,5-P$_3$ binding (*Protocol 9*) is done using a bovine adrenal binding protein (15), which is basically a crude P2 pellet of bovine adrenal cortex membranes. Bovine adrenal cortex is dissected out on ice and homogenized in 20 vol. of ice-cold 20 mM NaHCO$_3$ using a polytron (setting 6, three bursts of 15 sec). The preparations are then centrifuged at 1000 *g* for 10 min at 4 °C; the supernatant is removed and the pellet rehomogenized and centrifuged again. The combined supernatants are then centrifuged at 20 000 r.p.m. (38 000 *g*) at 4 °C for 20 min, and the membranes resuspended to 5–10 mg/ml protein in incubation buffer (20 mM NaCl, 100 mM KCl, 1 mM EDTA, 1 mg/ml BSA and 20 mM Tris–HCl, pH 8.3). Aliquots of this can be stored frozen. The binding assays are done on ice in a final volume of 250 μl, in Eppendorf tubes. Each assay contains 0.25–0.5 nM [^{3}H]-1,4,5-IP$_3$ (25 μl) and either cold 1,4,5-IP$_3$ (for doing displacement curves) or sample as appropriate (25 μl). Non-specific binding is defined by 10 μM cold 1,4,5-IP$_3$ (we use the Sigma inositol trisphosphate preparation). Reactions are started by addition of binding protein (150 μl) and samples are vortexed and incubated for 10 min. Incubations are terminated by centrifugation in a microfuge for 10 min; the supernatant is then poured off and the pellet washed twice with distilled water. The pellet is solubilized overnight with 100 μl tissue solubilizer, then 1 ml of scintillant is added and the samples counted. Tissue samples are prepared by one of the methods described earlier and 1,4,5-IP$_3$ levels are determined by measuring displacement of label and comparison to a displacement curve (0.1–100 nM) with standard 1,4,5-IP$_3$.

Inositol 1,3,4,5-P$_4$ binding (*Protocol 10*) is done using a rat cerebellar binding protein (17), which is basically a crude preparation of cerebellar membranes. Cerebella are removed from rats following decapitation and homogenized in 20 vol. of ice-cold 20 mM $NaHCO_3$/1 mM dithiothreitol. The preparations are then centrifuged twice at 20 000 r.p.m. (38 000 g) at 4 °C for 20 min, and the membranes resuspended to 5–10 mg/ml protein. Aliquots of this can be frozen. The binding assays are done on ice in a final volume of 160 µl in Eppendorf tubes. Each assay contains 25 mM sodium acetate, 25 mM K_2HPO_4, 2 mM EDTA (pH 5.0), approx. 15 000 d.p.m. of $[^{32}P]$-1,3,4,5-IP_4 and either cold 1,3,4,5-IP_4 (for doing displacement curves) or sample as appropriate. Non-specific binding is defined by 10 µM cold 1,3,4,5-IP_4. Reactions are started by addition of binding protein and samples are vortexed occasionally during the 30 min incubation period. Incubations are stopped by rapid filtering through Whatman GF/B filters and free ligand is removed by washing three times with 3 ml of 25 mM sodium acetate, 25 mM K_2HPO_4, 5 mM $NaHCO_3$ 1 mM EDTA (pH 5.0). Radioactivity remaining on the filters is determined by scintillation counting. Tissue samples are prepared by one of the methods described earlier and 1,3,4,5-IP_4 levels are determined by measuring displacement of label and comparison to a displacement curve with standard 1,3,4,5-IP_4.

6. Conclusions

In this chapter I have tried to cover most of the techniques used by workers involved in investigating the role played by inositol lipids and inositol phosphates in cellular signalling mechanisms. In such a large and expanding field, however, it is impossible to be cognisant with all the bewildering variety of techniques available, though I hope that I have provided sufficient information or references to cover most needs. While this chapter was in preparation a book covering a whole range of techniques in inositide research, including several not described here, was published (18). Those workers requiring a more in-depth analysis of these methods, including the ones mentioned in this chapter, may find that publication helpful.

Acknowledgements

I would like to thank all my technicians who helped with the experiments, particularly S. J. McClue, Z. Taghavi, and S. Neidhart. My thanks to Dr M. J. O. Wakelam for his help with the manuscript.

References

1. Berridge, M. J., Downes, C. P., and Hanley, M. R. (1982). *Biochem. J.*, **206**, 587–95.

2. Hajra A. K., Fisher S. K., and Agranoff B. W. (1988). In *Neuromethods*, Vol. 7 (ed. A. A. Boulton, G. B., Baker, and L. A. Horrocks), pp. 211–25. Humana Press, Clifton, NJ.
3. Dean N. M. and Beaven M. A. (1989). *Anal. Biochem.*, **183**, 199–209.
4. Palmer S. and Wakelam M. J. O. (1989). *Biochim. Biophys. Acta* **1014**, 239–46.
5. Wells M. A. and Dittmer J. C. (1965). *Biochemistry*, **4**, 2459–68.
6. Hawkins P. T., Berrie C. P., Morris A. J., and Downes C. P. (1987). *Biochem. J.*, **243**, 211–18.
7. Wreggett K. A. and Irvine R. F. (1987). *Biochem. J.*, **245**, 655–60.
8. Godfrey P. P. and Putney J. W. (1984). *Biochem J.* **218**, 187–95.
9. Creba J. A., Downes C. P. Hawkins P. T., Brewster, G., Michell R. H., and Kirk C. J. (1983). *Biochem. J.*, **212**, 733–47.
10. Godfrey P. P. (1989). *Biochem J.*, **258**, 621–4.
11. Leavitt A. L. and Sherman W. R. (1982). *Methods Enzymol.*, **89**, 3–18.
12. Kennedy E. D., Batty I. H., Chilvers E. R., and Nahorski S. R. (1989). *Biochem. J.*, **260**, 283–6.
13. Batty I., Letcher A. J., and Nahorski, S. R. (1989). *Biochem. J.*, **258**, 23–32.
14. Challiss R. A. J., Batty I., and Nahorski S. R. (1988). *Biochem Biophys. Res. Commun.*, **157**, 684–91.
15. Palmer S., Hughes K. T., Lee D. Y., and Wakelam M. J. O. (1989). *Cell. Signal.* **1**, 147–53.
16. Donie F. and Reiser G. (1989). *FEBS Lett.*, **254**, 155–8.
17. Challiss R. A. J. and Nahorski S. R. (1990). *J. Neurochem.*, **54**, 2138–41.
18. Irvine, R. F. (ed.) (1990). *Methods in inositide research*. Raven Press, New York.

6

Phosphatidylcholine hydrolysis by phospholipases C and D

R. W. BONSER and N. T. THOMPSON

1. Introduction

The importance of the receptor-coupled phosphatidylinositol 4,5-bis-phosphate (PIP_2)-specific phospholipase C (PLC) in intracellular signalling is now clearly established (for review see ref. 1). The products of this phospholipase have well-recognized second messenger roles. Inositol 1,4,5-trisphosphate (IP_3) mobilizes intracellular calcium and diacylglycerol (DAG) activates protein kinase C (PKC). The stimulated hydrolysis of PIP_2 by PLC has been regarded by many as an important source of DAG in activated cells. More recently, the source of the DAG that activates PKC has been questioned. A great deal of evidence now points to the receptor-linked hydrolysis of phosphatidylcholine (PC) as a major source of DAG in stimulated cells (for review see ref. 2). DAG can be released from PC by two separate mechanisms. The first is via the direct action of a PLC which generates DAG and phosphorylcholine as products. The other pathway involves the sequential actions of phospholipase D (PLD) and phosphatidate phosphohydrolase. Phosphatidic acid (PA) and choline constitute the initial products of PLD-dependent PC hydrolysis. Quantitating the formation of PA, DAG, choline, and phosphorylcholine, monitoring the kinetics of release and measuring product profiles can provide valuable information about which pathway(s) is operative in stimulated cells. Although the temporal relationship between PA and DAG formation is a useful indicator of PLC or PLD activation, it is important to note that both products can also be derived from other phospholipid pools, not just from PC, or by synthesis *de novo*. The first part of this chapter will describe some of the methods currently used to measure the products of PC hydrolysis. The second part will focus on the methods that exist for specifically measuring PLD and the techniques that can be applied to assess the contribution of PLC and PLD to phospholipid turnover in activated cells.

2. Choline and phosphorylcholine production

Two approaches have been used to monitor the release of choline and phosphorylcholine in activated cells. The first measures the release of these products from cells that have been pre-incubated with isotopically labelled choline. The other technique quantitates by radioenzymatic assay the amounts of choline and phosphorylcholine that are generated.

2.1 Pre-labelling cells with [³H]choline or [¹⁴C]choline

A variety of cells have been labelled in their choline phospholipid pools with high specific activity [methyl-³H]choline or [methyl-¹⁴C]choline, these include fibroblasts, endothelial cells, pancreatic islet cells, hepatocytes, and smooth muscle cells. Invariably the methods simply involve adding [methyl-³H]choline or [methyl-¹⁴C]choline to the culture medium (0.5–1 µCi/ml) and incubating the cells with the radiolabelled precursor (usually 18–48 h). The labelling medium is then removed and the cells are washed with a buffered, isotonic medium such as Hepes-buffered Hanks' balanced salt solution (HBH) or a Hepes-buffered serum-free growth medium. Cell activation can be carried out in the washing buffer or a medium of choice.

In other tissues, where prolonged incubation *in vitro* is unsuitable, the technique is limited by the pre-incubation period and the extent of precursor incorporation.

- It is advantageous to keep the volume of the incubation buffer during cell activation to a minimum (0.5–1 ml) since this minimizes the volumes of solvents that will subsequently be used to extract the choline-containing metabolites.

2.2 Extraction of choline-containing metabolites

Water-soluble choline-containing metabolites can be quantitatively extracted from tissues using a modification of procedures originally designed for isolating lipids (3).

Protocol 1. Extracting choline-containing metabolites from tissues

Materials
- chloroform
- methanol, cooled in dry ice
- screw-cap glass tubes with PFTE cap-liners (100 × 16 mm)

Method
1. Remove the incubation medium from control and stimulated cells and add to the extraction tubes.

2. Immediately afterwards add cold methanol (0.5–1 ml) to the cells to stop the reaction and place the culture dishes on ice for 10 min.

3. Scrape the cell monolayers from the dish and transfer to the extraction tubes.

4. Add chloroform so that the final ratio of chloroform:methanol:water is adjusted to 1:1:0.85 (v/v).

5. Cap the tubes and mix the contents thoroughly on a vortex mixer then centrifuge the tubes at 2000 *g* for 5 min to separate the emulsion into two phases.

6. Remove the upper aqueous–methanol phase by aspiration and analyse for choline-containing metabolites.

2.3 Separation of choline and phosphorylcholine

A number of procedures for separating choline from other choline-containing metabolites by ion-exchange chromatography have been described and the majority are based on the method originally developed by Dowdall *et al.* (4). A recent modification by Cook and Wakelam (5) efficiently separates choline and phosphorylcholine from glycerophosphocholine, the other major water-soluble choline-containing metabolite.

Protocol 2. Ion-exchange chromatography

Materials

- Dowex 50 WH$^+$ ion-exchange resin
- HCl
- Bio-Rad Econocolumns

Method

1. Wash the Dowex resin with 1 M HCl and then wash repeatedly with water until a stable pH (approx. 5.5) is reached.

2. Prepare small columns (1 ml packed volume) of the resin in Econo-columns or glass-wool-plugged Pasteur pipettes.

3. Dilute the aqueous–methanol phase from the chloroform/methanol extracted samples (see *Protocol 1*) to 5 ml with water and load this on to the column.

4. Wash with a further 4 ml of water; the fraction which is composed of the loading volume (5 ml) and the first water wash (4 ml) contains [methyl-^{3}H]glycerophosphocholine.

5. Elute the column with 20 ml 0.1 M HCl; this fraction contains phosphoryl[methyl-^{3}H]choline.

Protocol 2. *Continued*

6. Finally elute any [methyl-³H]choline with 20 ml of 1 M HCl.

7. Take 1 ml aliquots of each fraction and determine radioactivity by scintillation counting.

- The elution profile of the columns can be checked by monitoring the separation of radiolabelled choline, phosphorylcholine and glycero-phosphocholine standards using control columns (see *Figure 1*).

- Radiolabelled glycerophosphocholine is not commercially available but can be easily prepared by treating [*N*-methyl-³H]phosphatidylcholine (Amersham) with monomethylamine according to the method of Clarke and Dawson (6).

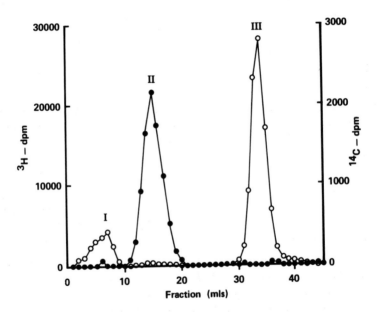

Figure 1. Separation of water-soluble choline-containing metabolites by ion-exchange chromatography. Isotopically labelled choline, phosphorylcholine, and glycerophospho-choline were resolved by ion-exchange chromatography (see *Protocol 2*). Peak I, [methyl-³H]glycerophosphocholine; Peak II, phosphoryl[methyl-¹⁴C]choline; Peak III, [methyl-³H]-choline.

Choline and choline-containing metabolites can also be readily separated by thin-layer chromatography (TLC). The most widely used procedure is based on the method of Yavin (7) as modified by Vance *et al.* (8).

Protocol 3. Thin-layer chromatography

Materials
- silica gel G TLC plates 20 × 20 cm (Whatman LK5D)
- methanol
- NaCl (0.9% w/v)
- 880 ammonia
- [methyl-^{14}C]choline and phosphoryl[methyl-^{14}C]choline (Amersham)

Method

1. Evaporate the aqueous–methanol fractions, from the chloroform/methanol extraction step (see *Protocol 1*) to dryness under a stream of nitrogen or in a Speed-vac evaporator (Savant).

2. Dissolve the dry residues in a small volume (< 50 μl) of ethanol:water (1:1, v/v) and spot each sample on to a separate lane of the TLC plate.

3. Spot radiolabelled choline and phosphorylcholine standards on to separate lanes of the TLC plate.

4. Develop the chromatogram in the solvent system methanol:0.9% NaCl:ammonia (50:50:5, v/v).

- Isotopically labelled choline and phosphorylcholine standards are located by autoradiography and the area of each lane of the TLC plate that corresponds to the position of each standard is scraped into a scintillation vial and counted.

- Alternatively, each lane of the chromatogram can be divided into 0.5-cm bands and each band of silica gel scraped and counted. The radioactivity profile of each lane can then be plotted and peaks of radiolabel that correspond to the position of the standards are identified and radioactivity quantitated.

- Both procedures are time-consuming and labour-intensive and are becoming less popular with the advent of the latest generation of TLC plate scanners (Berthold). These instruments will locate and quantitate peaks of radioactivity without the need for autoradiography, or scintillation counting.

2.4 Radioenzymatic assays for measuring levels of choline and phosphorylcholine

Some cells (for example, neutrophils) cannot be maintained in a viable state *in vitro* for long periods of time and therefore are unsuitable for the [methyl-^{3}H]choline pre-labelling experiments described above. In such circumstances the stimulated hydrolysis of PC can be monitored by measuring the amounts of choline and phosphorylcholine that are produced by radioenzymatic assay. Measuring the mass levels of these products offers several advantages over the [methyl-^{3}H]choline technique. First, as was alluded to earlier, the viability of the tissue is not compromised by extended pre-labelling periods. Second, hydrolysis of all the choline phospholipid pools can be monitored, not just those pools that rapidly incorporate [methyl-^{3}H]choline. Third, the amounts of choline or phosphorylcholine produced can be directly compared with measurements of diradylglycerol or phosphatidic acid production (see Sections 3.7 and 4.3).

2.5 Measurement of choline

Protocol 4. A radioenzymatic assay for measuring choline

Materials
- chloroform
- methanol
- formic acid
- buffer A = 25 mM Tris pH 9.0
- buffer B = 25 mM Tris pH 9.0, 10 mM MgSO$_4$
- buffer C = 25 mM Tris pH 10.0, 10 mM MgSO$_4$, 10 mM NaF, 2.5 mM EGTA,
- choline chloride
- phosphorylcholine (10 mg/ml in buffer B)
- γ-[^{32}P]ATP 20 mM in buffer B (specific activity = 50 d.p.m./pmol)
- Dowex 1-X4 ion-exchange resin
- choline kinase (Boehringer Mannheim)

Method
1. Suspend purified human neutrophils at 1 × 10^7 cells/ml in 5 mM Hepes-buffered Hanks' balanced salt solution containing 2.5 μM cytochalasin B and incubate at 37 °C for 10 min.
2. To 0.5 ml cells add the appropriate stimulus; for example, fMet–Leu–Phe or phorbol myristate acetate (PMA) dissolved in dimethylsulfoxide (DMSO), final concentration 0.03%, v/v and incubate at 37 °C for periods ranging from 0–300 sec.
3. Stop the reaction by adding 1.5 ml chloroform:methanol (2:1, v/v containing 20 mM formic acid).
4. Cap the extraction tubes, mix the contents thoroughly and centrifuge at 2000 g for 5 min to separate the emulsion into two phases.

5. Remove the upper aqueous–methanol fraction and evaporate to dryness under a stream of nitrogen or in a Speed-vac evaporator (Savant).

6. Dissolve the dry residue in 180 μl buffer C, add 10 μl of γ-[^{32}P]ATP reagent, 10 μl choline kinase (1 unit/ml in buffer A) and incubate at 37 °C for 2 h.

7. Stop the reaction by adding 10 μl of the phosphorylcholine reagent and 1.5 ml ice-cold buffer B. Stand all tubes on ice.

8. Wash the Dowex resin at least three times in buffer B and prepare columns (1.5 ml packed volume) in Econocolumns or glass-wool-plugged Pasteur pipettes.

9. Load the contents of the reaction tubes on to the columns and discard the eluant. Elute the [^{32}P]phosphorylcholine with 4.5 ml buffer B. Remove an aliquot of the eluant and measure radioactivity by scintillation counting.

● Standard curves are generated by adding known amounts of choline chloride to the assay and quantitating the amount of radioactivity incorporated into [^{32}P]phosphorylcholine. The amount of choline in each sample is then calculated by reference to the standard curve.

● The assay is linear in the range 10–1000 pmol choline chloride (see *Figure 2a*) and is specific for choline since any phosphorylated derivatives of ethanolamine and monomethyl- and dimethylethanolamine that are generated by choline kinase are not eluted from the ion-exchange resin (9).

● An example of the data obtained using this technique is shown in *Figure 2b*.

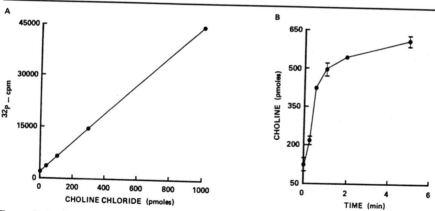

Figure 2. Radioenzymatic measurement of choline release.
A: Standard curve. 0–1000 pmol choline chloride were treated with choline kinase and [^{32}P]phosphorylcholine quantitated (see *Protocol 4*).
B: Time-course of choline release in fMet–Leu–Phe-stimulated human neutrophils. Purified neutrophils were treated with 100 nM fMet–Leu–Phe, extracted, and choline levels measured by radioenzymatic assay (see *Protocol 4*).

2.6 Measurement of phosphorylcholine

The assay for choline is easily modified to measure phosphorylcholine by treating the aqueous–methanol extract with alkaline phosphatase to hydrolyse phosphorylcholine to choline. The amount of phosphorylcholine in each sample is then calculated as the difference between the levels of choline measured before and after alkaline phosphatase treatment.

Protocol 5. A radioenzymatic assay for phosphorylcholine

Materials

- same as in *Protocol 4*
- alkaline phosphatase—bovine kidney affinity purified (Sigma)

Method

1. Extract choline-containing metabolites from stimulated human neutrophils according to the procedure described in *Protocol 4*, steps 1–5.

2. Evaporate the aqueous–methanol fraction to dryness under a stream of nitrogen or in a Speed-vac evaporator. Dissolve the dry residue in 340 μl buffer C and divide the solution equally into two.

3. One-half of the sample is treated according to the method described in *Protocol 4*, steps 6–9. The remainder is treated as in steps 4 and 5 below.

4. Add 10 μl alkaline phosphatase (5 units/ml in buffer C) and incubate at 37 °C for 30 min. Stop the reaction by heating the tubes in a boiling-water bath for 10 min. Stand all the tubes on ice.

5. Samples are then assayed for choline as described in *Protocol 4*, steps 6–9.

2.7 Alternative assays for phosphorylcholine

An alternative assay that directly measures the amount of phosphorylcholine in cells, originally developed by Choy *et al.* (10) and modified by Truett *et al.* (11), quantitates the conversion of phosphorylcholine to [^{32}P]CDP-choline by CTP-phosphorylcholine cytidylyltransferase in the presence of radiolabelled [α-^{32}P]CTP as substrate. Although this method provides an accurate measure of phosphorylcholine levels in cells, it is much less sensitive than the radioenzymatic assay for choline and only has a minimum detection limit of 1–20 nmol.

3. Phosphatidic acid

The production of phosphatidic acid (PA) can be monitored in intact cells by incorporating isotopically labelled precursors into the 'phosphatidyl' moiety of membrane phospholipids and measuring release of radiolabelled PA. The 'phosphatidyl' moiety of phospholipids can be radiolabelled principally by four methods:

(a) incorporating exogenous labelled fatty acids into cellular phospholipids

(b) providing radiolabelled glycerol as a precursor for phospholipid synthesis

(c) metabolically labelling the phosphate ester linkage of membrane phospholipids with [^{32}P]orthophosphate

(d) supplying radiolabelled lysophospholipids that are readily acylated by cells and incorporated into their phospholipid pools; for example, 1-O-[^{3}H]octadecyl-2-lysophosphatidylcholine (1-O-[^{3}H]octadecyl-2-lysoPC).

All four techniques have been used to generate information on PA formation in activated cells and a brief outline of the methods employed is provided below.

3.1 Fatty acid labelling

Exogenous radiolabelled arachidonate, oleate, palmitate, and myristate are rapidly incorporated into cellular phospholipids. Phospholipid labelling is readily achieved by adding isotopically labelled fatty acids to the growth medium of cells in culture (usually 1–10 µCi/ml). The specific radioactivity of the exogenous fatty acids can be greatly increased by eliminating serum from the culture medium. Under such circumstances it is advantageous to add bovine serum albumin to the medium (0.1–1 mg/ml). Arachidonic acid is present in serum at much lower concentrations than the other fatty acids and can be added directly to serum containing medium without appreciably reducing specific radioactivity. The incubation period varies amongst cell types but is usually 18–24 h.

3.2 Glycerol labelling

This procedure is less popular than fatty acid labelling but is equally simple. High specific activity [^{3}H]glycerol is added to serum-free culture medium (usually 5–25 µCi/ml) for periods ranging from 1.5–24 h. Unmetabolized [^{3}H]glycerol can easily be washed away before cells are treated with the appropriate stimulus.

3.3 Phosphate labelling

A major disadvantage of pre-labelling cellular phospholipids with [^{32}P]-orthophosphate is that in order to achieve a high specific activity, the incubation medium must be either phosphate-free or contain very low levels of inorganic phosphate. Cell viability can be seriously compromised under these conditions and extended incubation periods are not recommended. It is conventional to wash cells in a phosphate deficient medium before adding

[^{32}P]*ortho*phosphate (usually 20–250 µCi/ml). The cells are incubated for 2–4 h at 37 °C, then washed in an ice-cold isotonic buffer and finally resuspended in a phosphate-replete medium before the stimulus is added.

3.4 1-*O*-[^{3}H]octadecyl-2-lysoPC labelling

This lysophospholipid is rapidly acylated and incorporated into the choline phospholipid pool of some cells; for example, neutrophils and hepatocytes (12, 13). Cells are incubated with 1-*O*-[^{3}H]octadecyl-2-lysoPC (1–5 µCi/ml) for 0.5–4 h in a buffered medium containing 0.1–1% (w/v) bovine serum albumin, either with or without calcium. This technique has the advantage of selectively labelling the choline phospholipid pool, but its application is limited since not all cells incorporate this lysophospholipid into their membrane phospholipids.

3.5 1-*O*-[^{3}H]octadecyl-2-acyl-phosphatidic acid formation

---

Protocol 6. Measuring PA formation in human neutrophils

Materials

- chloroform
- methanol
- 1-*O*-[^{3}H]octadecyl-2-lysophosphatidylcholine (Amersham)
- buffer A = 25 mM Hepes pH 7.2, 125 mM NaCl, 10 mM glucose, 1 mM EGTA, bovine serum albumin (1 mg/ml)

Method

1. Evaporate 1-*O*-[^{3}H]octadecyl-2-lysoPC to dryness under a stream of nitrogen and dissolve in buffer A at 1 µCi/ml.

2. Suspend purified human neutrophils at 2×10^7 cells/ml in this pre-labelling buffer and incubate at 37 °C for 30 min.

3. Pellet the cells by centrifuging at 200 *g* and wash the pre-labelled neutrophils twice in ice-cold Hepes-buffered Hanks' balanced salt solution (HBH); finally resuspend at 2×10^7 cells/ml in HBH.

4. To 0.5 ml cell suspension add 2.5 µl cytochalasin B (1 mM in DMSO) and incubate the neutrophils at 37 °C for 10 min.

5. Add the appropriate agonist; for example, fMet–Leu–Phe or PMA, dissolved in either DMSO or ethanol and incubate for varying periods (0– 300 sec) at 37 °C.

6. Stop the reaction by adding 1.5 ml chloroform/methanol (1:2, v/v). Extract total lipids by adding 0.5 ml 1 M NaCl and a further 0.5 ml chloroform and mix thoroughly on a vortex mixer.

7. Centrifuge the extraction tubes at 2000 *g* for 5 min to separate the emulsion into two phases and remove the lower chloroform layer.
8. Separate PA from other lipids by TLC (see *Protocol 7*).

3.6 Isolation of phosphatidic acid by TLC

Phosphatidic acid can be easily separated from other lipid species by TLC. A number of one- and two-dimensional systems have been developed (14–17). A simple and effective one-dimensional system (see *Protocol 7*) has been routinely used by our group to isolate PA (18).

Protocol 7. Separating PA from other lipids by silica gel TLC

Materials
- silica gel G TLC plates 20 × 20 cm (Whatman LK5D)
- chloroform
- methanol
- propan-1-ol
- 2,2,4-trimethylpentane
- ethyl acetate
- glacial acetic acid
- pyridine
- 70% formic acid
- KCl (0.25%, w/v)
- [glycerol-^{14}C (U)]phosphatidic acid (NEN)

Method
1. Evaporate the total lipid extract (see *Protocol 6*, step 7) to dryness under a stream of nitrogen and dissolve the dry lipid film in a small volume (50–100 μl) of chloroform/methanol (95:5, v/v).
2. Apply the extracted lipid samples to individual lanes of the TLC plate and spot the [^{14}C]PA standard on to a separate lane.
3. Develop the chromatogram in the solvent system chloroform:methanol:propan-1-ol:0.25% KCl:ethyl acetate (25:13:25:9:25, v/v).
4. Allow the chromatogram to dry thoroughly and scrape regions of the TLC plate that correspond to the position of the PA standard into scintillation vials, add 4 ml scintillant and count.
- The 1-*O*-[^{3}H]octadecyl-PA produced is located by comparing the migration of each radiolabelled lipid with the authentic PA standard.

Protocol 7. *Continued*

- Radiolabelled lipids can be located by autoradiography or by scraping the TLC plate as described in *Protocol 3*. Alternatively, use a TLC plate scanner (Berthold) to locate and quantitate tritium-labelled lipids.

- The homogenity of the PA fraction must be confirmed by comparing its migration with the authentic lipid standard in at least one other solvent system; for example, ethyl acetate:2,2,4-trimethylpentane:glacial acetic acid (9:5:2, v/v) or chloroform:pyridine:70% formic acid (50:30:7, v/v).

3.7 Quantitating levels of phosphatidic acid in cells

Techniques that were originally developed to quantitate total lipid phosphorus have been applied to the measurement of PA. Measuring lipid phosphorus requires digestion with strong acids and oxidizing agents and a colorimetric estimation of the inorganic phosphate that remains (19, 20). These are laborious techniques that are not ideally suited to quantitating the nanomolar amounts of PA generated by cells, unless all phosphate-containing impurities from samples, glassware and reagents are removed first. Recently, two alternative procedures for quantitating phospholipids have been developed and these have been used to measure PA production in hepatocytes, neutrophils and sea urchin spermatozoa (21–24).

3.8 Charring method

This procedure chars any lipid species that can be resolved by TLC and quantitates the amount of lipid in each spot by densitometric analysis. Although this procedure appears straightforward, experience has demonstrated that considerable practice is required to obtain reproducible and reliable results.

Protocol 8. Quantitating PA formation by charring and densitometry

Materials

- copper sulfate
- phosphoric acid
- phosphatidic acid, egg yolk (Avanti)

Method

1. Separate PA from other lipid species by TLC (see *Protocol 7*) and dry the chromatogram thoroughly.

2. Dip the TLC plate into a solution of $CuSO_4$ (10% w/v) and phosphoric acid (8% w/v), remove and again ensure the plate is dried thoroughly.

3. Heat the plate at 185 °C for 25–30 min in an oven to char the lipids and then scan the plate with a laser densitometer (LKB Ultroscan) as described by Goppelt & Resch (25).

- Standard curves are generated by spotting known amounts of authentic PA (1–50 μg) on to the chromatogram before developing it in the appropriate solvent system. The chromatogram is then charred according the procedure described above and the density of each spot measured.

- This method measures PA levels as low as 1 μg (approx. 1.3 nmol) and is best suited for the analysis of large amounts of tissue.

3.9 Dye staining

3.9.1 Coomassie blue staining

This method was originally developed by Nakamura and Hanada (26) and has been used to measure PA in lipid extracts from neutrophils and hepatocytes (22, 23). This technique is similar to the charring method (see *Protocol 8*) and also uses densitometry to quantitate lipids separated by TLC. A great deal of practical experience is needed to ensure reproducible results

---

Protocol 9. Quantitating PA by Coomassie blue staining

Materials

- Coomassie blue R250 (Aldrich)
- methanol
- NaCl
- phosphatidic acid, egg yolk (Avanti)

Method

1. Separate PA from other lipids by TLC using the procedure described in *Protocol 7* and dry the chromatogram thoroughly.

2. Immerse the TLC plate for 30 min in a solution of 30% methanol containing Coomassie blue (0.03%, w/v) and 100 mM NaCl. Then destain the silica gel plate by transferring it to a solution of 30% methanol containing 100 mM NaCl for 5 min.

3. Dry the chromatogram thoroughly and scan the bands on the plate with a laser densitometer at a wavelength of 633 nm.

- Like the charring method, standard curves are produced by spotting known amounts of an authentic PA standard onto the chromatogram before development.

Protocol 9. *Continued*

- The Coomassie blue staining procedure is more sensitive than the charring method and is reported to detect levels of PA as low as 250 ng (approx. 300 pmol).

3.9.2 Victoria blue staining

An alternative dye staining method has been described by Eryomin and Poznyakov (27) that does not require the staining and destaining of TLC plates. Quantitation is based on a spectrophotometric measurement instead of laser densitometry.

Protocol 10. Measuring PA formation by Victoria blue staining

Materials

- hexane
- isopropanol
- ammonium hydroxide (1.5 M)
- Victoria blue (Aldrich)
- chloroform
- colorimetric reagent = Victoria blue, 0.0135% w/v dissolved in ethylene glycol:glycerol (1:1, v/v)

Method

1. Separate lipids by TLC, locate and identify the PA bands, as described in *Protocol 7*, or by iodine staining.
2. Scrape the area of silica gel corresponding to the position of the PA standard into a glass tube.
3. Add 200 μl hexane:isopropanol:1.5 M NH$_4$OH (24:16:1, v/v), 0.5 ml of the colorimetric reagent and 2.8 ml chloroform.
4. Mix thoroughly, add 1 ml of ethylene glycol, mix again and then centrifuge the tubes at 2000 g to separate the two phases.
5. Remove the lower chloroform layer and measure the absorbance at 590 nm in a spectrophotometer.

- Standards and blanks are treated the same way and standard curves generated.
- This method is much less sensitive than the Coomassie blue stain or charring techniques and only measures PA levels above 10–20 μg (approximately 15–30 nmol).

- Interference from contaminants such as detergents (for example, SDS) or buffers (for example, Tris and Hepes) also limits the utility of this method, although washing procedures can reduce background contamination (27).

---

4. Diradylglycerols

Production of diradylglycerols* (DRG) in activated cells can be monitored by isotopically labelling phospholipid pools and separating the radiolabelled products formed by TLC. Alternatively, DRG can be quantified by radioenzymatic assay or by densitometric analysis of the resolved DRG bands following staining or charring of the chromatogram. Each method has its advantages and disadvantages which will be addressed later.

4.1 Isotopic labelling of phospholipid pools

A number of isotopically labelled precursors can be used to radiolabel the 'phosphatidyl' moiety of phospholipid pools and general procedures for their use have been described in 3.1–3.4. Precursors that have been used routinely to monitor DRG production in activated cells include, isotopically labelled free fatty acids (FFA), glycerol, [^{3}H]hexadecanol and 1-O-[^{3}H]acyl- or 1-O-[^{3}H]alkyl-2-lysophosphatidylcholine (12, 28–31). Careful choice of the isotopically labelled precursor can provide information on the source of the DRG and its likely fatty acid composition. The choice of precursor will depend on the tissue and the phospholipid pool being investigated. Important points to remember in making the choice are:

(a) Some precursors may be extensively metabolized in certain tissues, either during the labelling period or following cell activation. This is a particular problem with arachidonic acid, and care should be taken to ensure that the TLC separation procedures can resolve labelled DRG from labelled eicosanoids.

(b) For non-cultured cells and intact tissues, where short pre-labelling periods are desirable, choose a precursor that is rapidly incorporated; for example, FFA. Glycerol and hexadecanol may not be suitable under such circumstances.

(c) In some tissues PC can be preferentially labelled with a saturated fatty acid; for example, palmitate or myristate (32, 33).

4.2 Isolation of cellular DRG by TLC

Numerous TLC systems have been described which will resolve radiolabelled DRG from other lipids. The system best suited to a particular application

* The term 'radyl' refers to acyl, alkyl, and alkenyl substituents.

Protocol 11. Separating DAG from other lipid species by TLC

Materials

- chloroform
- methanol
- benzene
- silica gel G TLC plates 20 × 20 cm (Whatman LK5D)
- 1-stearoyl-2-[1-^{14}C]arachidonyl-*sn*-glycerol ([^{14}C]SAG) (Amersham)

Method

1. Extract total lipids from activated cells and apply the chloroform extracts and the [^{14}C]SAG standard to separate lanes of the TLC plate (see *Protocols 6* and *7*).

2. Develop the chromatogram in the solvent system benzene:chloroform: methanol (80:15:5, v/v). Remove the plate from the chromatography tank and dry thoroughly.

3. Locate the position of the [^{14}C]SAG standard either by autoradiography or by scraping bands of silica (0.5 cm wide) from the TLC lane and quantitate radioactivity by scintillation counting (see *Protocol 3*).

4. Scrape regions of the chromatogram that correspond to the position of the [^{14}C]SAG standard into scintillation vials, add 4 ml scintillant, and count.

---

depends on the contaminating lipid species present. The procedure described above (30) is useful for separating DRG ($R_f = 0.39$) from monoacylglycerol ($R_f = 0.07$) and triacylglycerol ($R_f = 0.79$).

4.3 Quantitation of DRG by radioenzymatic conversion to [^{32}P]phosphatidic acid

The enzymatic conversion of DRG to [^{32}P]PA by diacylglycerol kinase (DGK) is a very sensitive means of quantitating DRG levels in cells. The method described below is based on that of Preiss *et al.* (34) and uses a commercially available DGK preparation from *Escherichia coli*.

Protocol 12. A radioenzymatic assay for DRG

Materials

- diacylglycerol kinase—*E. coli* (Lipidex Inc.)
- phosphatidylserine (PS) from bovine brain (0.165 mg/ml) dissolved in chloroform:methanol, (95:5, v/v) (Lipid Products) γ-[^{32}P]ATP (Amersham)

- Triton X-100 (1.5%, w/v in buffer 1)
- dithiothreitol (20 mM in buffer 1)
- perchloric acid (1% w/v)
- 1-stearoyl-2-arachidonyl-*sn*-glycerol (Avanti)
- 1-stearoyl-2-[1-^{14}C]arachidonyl-*sn*-glycerol (Amersham)
- silica gel G TLC plates 20 × 20 cm (Whatman LK5D)
- buffer 1 = 100 mM imidazole pH 6.6, 1 mM diethylenetriamine-penta acetic acid (DTPA)
- buffer 2 = 100 mM imidazole pH 6.6, 100 mM NaCl, 25 mM MgCl$_2$, 2 mM EGTA

Method

1. Add 100 µl of the PS solution to the chloroform extracts (see *Protocol 6*) and evaporate to dryness under streams of nitrogen.
2. Prepare standard curves by adding 50, 100, 200, and 500 pmol of SAG together with 100 µl of the PS reagent to glass tubes and evaporate to dryness under nitrogen.
3. Add 20 µl of Triton X-100 solution and 50 µl of buffer 2; mix on a vortex mixer for 5–10 sec, then leave to stand for 10 min.
4. Mix again on a vortex mixer for a further 10 sec, then agitate the tubes in a sonicating water bath for 5–10 sec. Repeat this step to ensure that all the lipid is dispersed in a micellar form.
5. To each tube add 10 µl of freshly prepared dithiothreitol reagent, 10 µl DGK (0.33 units/ml in buffer 2) and 10 µl 10 mM γ-[^{32}P]ATP (specific activity = 50 µCi/µmol dissolved in buffer 1), mix thoroughly and incubate at 30 °C for 1 h.
6. Stop the reaction by adding 1 ml 1% perchloric acid. Extract lipids by adding 2.4 ml chloroform:methanol (1:1, v/v), mix thoroughly on a vortex mixer and centrifuge the tubes at 2000 *g* for 5 min to separate the emulsion into two phases.
7. Remove and discard the upper aqueous–methanol phases (**care!** these contain the bulk of the γ-[^{32}P]ATP). Wash the lower chloroform layers by adding 2 ml of theoretical upper phase (the aqueous phase of chloroform:methanol:1% perchloric acid (6:6:5, v/v)), mix, and then separate phases by centrifugation.
8. Remove the lower chloroform layers and evaporate to dryness under streams of nitrogen. Dissolve the dry lipid extracts in a small volume (50–100 µl) of chloroform:methanol (95:5, v/v) and apply to separate lanes of the TLC plate.

Protocol 12. *Continued*

9. Develop the chromatogram in the solvent system chloroform:methanol: glacial acetic acid (65:15:5, v/v). Remove the plates, dry thoroughly, locate [^{32}P]PA by autoradiography and scrape bands of silica corresponding to the position of the [^{32}P]PA ($R_f = 0.54$–0.84) into scintillation vials and count.

- SAG is unstable and should be stored dry at -30 °C under nitrogen. Experience has also shown that the most reliable means of quantitating the conversion of DRG to PA and measuring the recovery of this product is to use known amounts of [^{14}C]SAG as a standard. 50–100 pmol of [^{14}C]SAG are analysed by TLC before and after treatment with DGK and radioactivity in the DRG and PA fractions quantitated. This provides information about the extent of DRG conversion to PA and the overall recovery of PA in the assay. Standard curves prepared using unlabelled SAG are still required to ensure that enzymatic conversion is linear over the required concentration range.

- The method described in Preiss *et al.* (34) uses a mixed micelle support composed of cardiolipin/octylglucoside. The modification to the assay described above is more efficient at dispersing lipid extracts into detergent micelles.

- Amounts of DRG in each sample are calculated by reference to the standard curve (see *Figure 3a*). The conversion of DRG to PA and the recovery of the PA after TLC (monitored using the [^{14}C]SAG standard) should always be > 90% and linear in the range 10–1000 pmol.

- The DGK assay is clearly time-consuming; however, this disadvantage is outweighed by its sensitivity, and in our experience the assay will reproducibly measure DRG levels down to 10 pmol.

- An example of the data obtained using this assay is shown in *Figure 3b*.

4.4 Quantitation of DRG by high-performance liquid chromatography

High-performance liquid chromatography (HPLC) is a powerful technique for quantitating different lipid classes, including DRG. A method specifically for measuring DRG has been described that uses refractive index detectors to monitor HPLC column eluants. However, HPLC has a minimum detection level of only 2–10 μmol and cannot resolve 1,2 diacyl-*sn*-glycerols from 2,3 diacyl-*sn*-glycerols and 1,3 diacyl-*sn*-glycerols. A complete description of this HPLC technique is beyond the scope of this chapter and readers are referred to the original paper for more details (35).

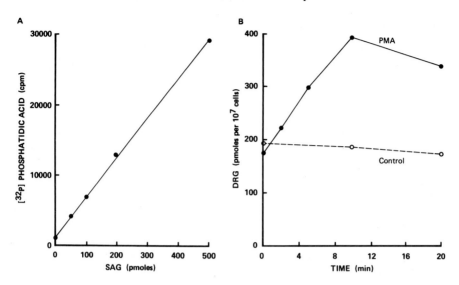

Figure 3. Quantitation of DRG using *E. coli* DGK.
A: Standard curve. 0–500 pmol SAG were treated with DGK and [32P]PA quantitated as described in *Protocol 12*. Values represent a recovery of > 92% for each SAG concentration (specific activity of γ[32P]ATP = 61880 c.p.m./μmol).
B: Time-course of PMA-stimulated DRG formation in HL60 granulocytes. DMSO-differentiated HL60 cells were treated with 100 nM PMA for various times and total DRG measured (see *Protocol 12*). DRG was quantitated by comparison with the standard curve shown in **A**.

4.5 Quantitation of DRG using charring and Coomassie blue staining methods

The methods described in Sections 3.8 and 3.9 for measuring PA by charring or Coomassie blue staining of thin-layer chromatograms can also be used to quantitate DRG levels. The techniques differ only in that lipid extracts should be chromatographed in a system that will resolve DRG (see *Protocol 11*) prior to charring or staining (see *Protocols 8* and *9*). Both methods are less sensitive than the DGK assay and have minimum detection limits of approximately 1 nmol. Considerable practical experience is required to ensure reliable and reproducible quantitation by densitometry.

4.6 Distinguishing between 1-acyl-, 1-*O*-alkyl-, and 1-*O*-alk-1′-enyl glycerides

All three lipid species can be produced in stimulated cells and it is important to distinguish between them when considering their possible roles as second messengers. None of the methods described so far will make any distinction between substituents at the C1 position of glycerides. There are, however, three techniques which can be used in conjunction with previously described methods to help determine the level of each individual species. Each procedure is based on the susceptibility of the different lipid species to chemical or enzymatic degradation. Analysis of the DRG fraction after chemical or enzymic treatment can provide some measure of the contribution each species makes to the total cellular DRG pool.

4.6.1 Alkaline hydrolysis of 1-acyl-species

To carry out this procedure simply dissolve the dry lipid extract (see *Protocols 6 and 7*) in 0.25 ml chloroform and mix with 0.25 ml of 0.2 mM NaOH in methanol. Incubate at room temperature for 1 h and then extract lipids by adding 0.5 ml of 1 M NaCl, mix thoroughly on a vortex mixer and separate phases by centrifugation at 2000 g for 5 min. Remove the lower chloroform phase containing the remaining DRG species and analyse this fraction by TLC. Diacylglycerol species are hydrolysed to free fatty acid and glycerol whilst 1-*O*-alkyl-2-acyl-*sn*-glycerols undergo partial hydrolysis to form 1-*O*-alkyl-2-lyso-*sn*-glycerols. It is therefore important that the TLC system used to separate the hydrolysis products can resolve monoglyceride species (see Section 4.2 and *Protocol 11*). Monoglycerides are substrates for DGK and are phosphorylated to form [^{32}P]lysoPA. This product migrates very close to other radiolabelled lipids generated in the DGK assay (for example, ceramide phosphate) and thus direct quantitation of monoglycerides in the hydrolysate is difficult. Nevertheless, the DGK assay can be used in conjunction with alkaline hydrolysis to measure amounts of ether-linked lipids in the total lipid pool.

First, isolate the [^{32}P]PA formed in the DGK assay by TLC and scrape the PA band from the chromatogram (see *Protocol 12*). Second, treat the silica gel with methanolic NaOH and extract the hydrolysis products as described above. Finally, separate 1-*O*-alkyl-2-lyso[^{32}P]PA from other lipids by TLC (see *Protocol 12*, R_f = 0.2) and measure radioactivity.

4.6.2 Removal of 1-*O*-alkl-1′-enyl species with HgCl$_2$

For this method dissolve lipid extracts in 0.3 ml chloroform:methanol (1:2, v/v), add 8 μl 50 mM HgCl$_2$, mix, and incubate at 37 °C for 30 min. Extract lipids by adding 0.2 ml of 1 M NaCl and 0.1 ml chloroform, mix thoroughly and separate phases by centrifugation. Analyse the lower chloroform layer for residual DRG species by TLC (see *Protocol 11*) or quantitate DRG using the DGK assay (see *Protocol 12*).

4.6.3 Sensitivity to *Rhizopus arrhizus* lipase

The phospholipase A$_1$ (PLA$_1$) activity of *R. arrhizus* lipase hydrolyses the 1-acyl moiety but not the 1-*O*-alkyl or 1-*O*-alk-1′-enyl substituents of phospholipids. A procedure described by Tyagi *et al.* (36) uses the purified lipase to hydrolyse the [^{32}P]PA formed by DGK (see *Protocol 12*) and thus selectively remove 1-acyl-PA species. This method has the added advantage over the alkaline hydrolysis procedure (see Section 4.6.1) in that additional extraction and tlc procedures are not required. The PLA$_1$ activity of the *R. arrhizus* lipase is specific for phospholipids and will not hydrolyse glycerides.

Protocol 13. Quantitating the amount of 1,2 diacyl-*sn*-glycerol in the total DRG pool

Materials

- same as in *Protocol 12*
- Rhizopus lipase (Boehringer Mannheim)

Method

1. Extract lipids from stimulated cells (see *Protocol 6*) and add 100 μl of the PS solution to the chloroform fraction. Evaporate lipids to dryness under a stream of nitrogen.

2. Treat each sample according to the procedures described in *Protocol 12*, steps 3 to 5.

3. Split the reaction mixture equally into two. Add 0.5 ml of 1% perchloric acid to one-half of the sample and extract total lipids by adding 1.2 ml chloroform:methanol (1:1, v/v), mix thoroughly, and separate phases by centrifugation. Wash the lower chloroform layer with 1 ml of theoretical upper phase (see *Protocol 12*, step 7) and evaporate the washed chloroform fraction to dryness under nitrogen.

4. To the other half of the sample add 500 units of Rhizopus lipase (50 000 units/ml) and incubate for 2 h at 30 °C. Stop the reaction by adding 0.5 ml of 1% perchloric acid and extract total lipids as described in step 3.

5. Separate [^{32}P]PA from other lipids by TLC according to the method described in *Protocol 12*, steps 8 and 9, and quantitate the amounts of DRG in both halves of the sample.

- The difference between the amounts of DRG measured before and after Rhizopus lipase treatment provides a measure of the DRG that was present as the 1-acyl species.

---

5. Measuring PLD activity via transphosphatidylation reactions

Analysing the kinetics of choline vs. phosphorylcholine and PA vs. DRG production can provide useful information about the contribution of the PLC and PLD pathways to agonist-stimulated PC breakdown in cells. However, the interconversion of these products by kinases and phosphohydrolases limits the interpretations that can be put on the data. More conclusive evidence, particularly on the importance of the PLD pathway, has come from the study of transphosphatidylation reactions in cells.

Transphosphatidylation is a unique PLD-dependent reaction that transfers the 'phosphatidyl' moiety of a phospholipid to a nucleophilic acceptor. When water is the acceptor the product formed is PA. In the presence of an aliphatic alcohol (for example, ethanol, propanol or butanol) the 'phosphatidyl' moiety is transferred to the alcohol to produce phosphatidylalcohols. These unnatural products are metabolically stable and accumulate in cells, thus providing a valuable measure of PLD activation. Transphosphatidylation reactions can be measured by metabolically labelling the 'phosphatidyl' moiety of membrane phospholipids with isotopically labelled precursors; for example, fatty acids, glycerol, orthophosphate, or radiolabelled lysophospholipids as described in Sections 3.1–3.5. Pre-labelled cells are then stimulated in the presence of high concentrations of the alcohols (30–200 mM) and the radiolabelled phosphatidylalcohols formed are isolated and quantitated (37). Alternatively, cells can be activated in the presence of a radiolabelled alcohol; for example, [^{3}H]butan-1-ol and the formation of [^{3}H]phosphatidyl-butanol (PBut) measured (38).

5.1 Pre-labelling technique

Protocol 14. Measuring transphosphatidylation reactions in neutrophils prelabelled with 1-*O*-[^{3}H]octadecyl-2-lysoPC

Materials
- silica gel G TLC plates 20 × 20 cm (Whatman LK5D)
- chloroform
- methanol
- 2,2,4-trimethylpentane
- ethyl acetate
- glacial acetic acid
- butan-1-ol
- 1-*O*-[^{3}H]octadecyl-2-lysophosphatidylcholine (Amersham)

Method
1. Pre-label purified human neutrophils with 1-*O*-[^{3}H]octadecyl-2-lysoPC as described in *Protocol 6*.
2. Wash the pre-labelled cells twice in ice-cold Hepes-buffered Hanks' balanced salt solution (HBH) and resuspend at 4 × 10^7 cells/ml in HBH.

3. To 0.25 ml cells add 0.25 ml HBH containing 20 mM butan-1-ol and 10 μM cytochalasin B, then incubate at 37 °C for 5 min.

4. Start the reaction by adding the appropriate stimulus (dissolved in DMSO, final concentration 0.03%, v/v) and incubate for 0–300 sec at 37 °C. Stop the reaction by adding 1.5 ml chloroform:methanol (1:2, v/v).

5. Extract total lipids by adding 0.5 ml 1 M NaCl and a further 0.5 ml chloroform and mix thoroughly on a vortex mixer. Spin the extraction tubes at 2000 g for 5 min to separate the emulsion into two phases and remove the lower chloroform layer.

6. Evaporate the chloroform extract to dryness under a stream of nitrogen and redissolve in a small volume (50–100 μl) of chloroform:methanol (95:5, v/v).

7. Apply the lipid samples to individual lanes of the TLC plate and spot an authentic [^{14}C]PBut standard (see *Protocol 15*) on to a separate lane.

8. Develop the chromatogram in a solvent system comprising the organic phase of 2,2,4-trimethylpentane:ethyl acetate:acetic acid:water (50:110: 20:100, v/v)

9. Locate areas of the chromatogram that correspond to the position of the [^{14}C]PBut standard (see *Protocol 3*), scrape the silica gel into scintillation vials, add 4 ml scintillant, and count.

Protocol 15. Preparation of a [^{14}C]PBut standard

Materials

- 1-stearoyl-2-[^{14}C]arachidonyl phosphatidylcholine (Amersham)
- phospholipase D–cabbage (Sigma)
- sodium dodecyl sulfate (SDS)
- butan-1-ol
- chloroform
- methanol
- buffer A = 0.1 M sodium acetate pH 5.6

Method

1. Place 1 μCi (17.86 μmol) 1-stearoyl-2-[^{14}C]arachidonyl phosphatidyl- choline in a glass tube and remove solvents under a stream of nitrogen.

2. Disperse the dry lipid in 100 μl of SDS (5 mM in buffer A) by sonicating briefly in a bath sonicator. Add 100 μl of CaCl$_2$ (375 mM in buffer A), 75 μl butan-1-ol and 625 μl buffer A.

Protocol 15. *Continued*

3. Start the reaction by adding 100 units of cabbage phospholipase D (1000 units/ml in buffer A) and incubate at 30 °C for 3 h.

4. Stop reaction by adding 3 ml of chloroform:methanol (1:2, v/v) and 1 ml of 1 M NaCl. Add a further 1 ml of chloroform, mix thoroughly, centrifuge at 2000 *g* to separate the emulsion into two phases, and remove the lower chloroform layer.

5. Evaporate the chloroform extract to dryness under a stream of nitrogen, dissolve the dry lipid in 100 μl of chloroform:methanol (95:5, v/v), spot 10% on to one lane of the TLC plate and the remainder on to a separate lane.

6. Develop the chromatogram in a solvent system comprising the organic phase of 2,2,4-trimethylpentane:ethyl acetate:acetic acid:water (50:110: 20:100, v/v); running time approx. 90 min.

7. Locate the [^{14}C]PBut formed ($R_f = 0.36 \pm 0.06$) by dividing the lane of the TLC plate on which 10% of the sample was applied, into 0.5-cm bands; scrape the bands into scintillation vials, add 4 ml scintillant, and count.

8. From the radioactivity profile of the chromatogram determine the exact location of the bulk of the [^{14}C]PBut, scrape the silica gel into a glass tube and elute four times with 1 ml chloroform:methanol (1:2, v/v). Store the [^{14}C]PBut standard as as dry lipid film under anhydrous conditions at −20 °C.

5.2 [^{3}H]butan-1-ol technique

The use of [^{3}H]butan-1-ol to monitor PLD activation has a number of advantages over the pre-labelling method (see *Protocol 14*). First, the pre-labelling of cells with radioactive phospholipid precursors is avoided. This is particularly important when the precursor may be converted to a biologically active product as with, for example, 1-*O*-octadecyl-2-lysoPC (lyso-platelet activating factor) or arachidonic acid. Furthermore, tissue viability is no longer compromised by extended pre-incubation periods. Problems concerning labelling to isotopic equilibrium are eliminated and the formation of PBut from *all* phospholipid substrates can be measured, not just those pools that are radiolabelled with a suitable precursor. Second, the high specific activity of the [^{3}H]butan-1-ol ensures that only very low concentrations need to be used; i.e. levels that are 1000 times less than those used in other procedures. The advantages are threefold: (*i*) only a small percentage of the phospholipid substrate(s) is converted to [^{3}H]PBut and hence total phospholipid metabolism through the PLD pathway is not substantially altered; (*ii*) the possibility of cellular toxicity from aliphatic alcohols is reduced substantially;

and (*iii*) the chemical stability of stimulants (either particulate or soluble) is not compromised.

Protocol 16. Monitoring PLD activity by measuring the incorporation of [³H]butan-1-ol into [³H]PBut

Materials

- [³H]butan-1-ol (27.8 Ci/mmol) (Amersham)
- chloroform
- methanol
- silica gel G TLC plates 20 × 20 cm (Whatman LK5D)
- 2,2,4-trimethylpentane
- ethyl acetate
- glacial acetic acid

Method

1. Purify human peripheral blood neutrophils and suspend at 4×10^7 cells/ ml in Hepes-buffered Hanks' balanced salt solution (HBH).
2. To 0.25 ml cell suspension add 0.25 ml HBH containing cytochalasin B (10 μM) and [³H]butan-1-ol (40 μCi/ml) and incubate at 37 °C for 5 min.
3. Start the reaction by adding the appropriate stimulus (dissolved in DMSO, final concentration 0.03%, v/v) and incubate at 37 °C for 0–300 sec.
4. Stop the reaction by adding 2 ml ice-cold HBH and stand all tubes on ice. Pellet the cells by centrifuging at 200 g for 2 min (4 °C) and discard the supernatants.
5. Add 1.5 ml chloroform:methanol (1:2, v/v) and 0.5 ml HBH to the cell pellet. Extract total lipids by adding a further 0.5 ml chloroform and 0.5 ml 1 M NaCl, then mix thoroughly on a vortex mixer and separate the emulsion into two phases by centrifuging at 200 g for 5 min.
6. Remove the lower chloroform layer and evaporate to dryness under a stream of nitrogen.

   NB **All procedures up to and including this step should be performed in a fume cupboard to minimize exposure to [³H]butan-1-ol vapour! Subsequent steps can be performed at the bench.**
7. Dissolve the dry lipid extract in 1 ml chloroform and wash three times with 2 ml theoretical upper phase [aqueous phase of chloroform: methanol:1 M NaCl:water (2:2:1:1, v/v)]. Transfer the washed chloroform fraction to a glass tube and evaporate to dryness in a stream of nitrogen.

Protocol 16. *Continued*

8. Redissolve the lipid extract in 50–100 μl chloroform:methanol (95:5, v/v) and spot on to a silica gel G TLC plate. Isolate [³H]PBut by TLC using the procedures described in *Protocol 14* and quantitate radioactivity by scintillation counting (see *Figure 4*).

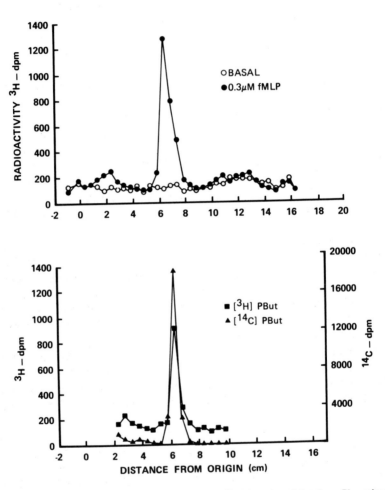

Figure 4. Incorporation of [³H]butan-1-ol into [³H]PBut by fMet–Leu–Phe-stimulated human neutrophils. Neutrophils were incubated with 0.3 μM fMet–Leu–Phe (●) or vehicle (○) for 5 min at 37 °C in the presence of [³H]butan-1-ol according to *Protocol 16*. The [³H]PBut formed was separated by TLC and radioactivity quantitated as described in *Protocol 14*. The lower panel shows the TLC profile of a PBut standard prepared by incubating 1-stearoyl-2-[¹⁴C]arachidonyl phosphatidylcholine with cabbage phospholipase D in the presence of [³H]butan-1-ol (see *Protocol 15*). ■, ³H d.p.m.; ▲, ¹⁴C d.p.m.

6. Phospholipase C

Phospholipase C does not catalyse a specific reaction analogous to PLD-dependent transphosphatidylation, consequently a direct and selective assay for PLC activation in intact cells is presently not available. The difficulty in selectively measuring PLC is compounded further by the fact that potent and selective inhibitors of PLC and PLD have yet to be identified. Nevertheless, some information on the contribution of PLC to PC breakdown has been obtained by exploiting the transphosphatidylation property of PLD. Trans-phosphatidylation has already been described in detail (see Section 5) but an important feature of this unique PLD-dependent reaction is the production of phosphatidylalcohols at the expense of PA (37, 39). Aliphatic alcohols can therefore be used as inhibitors of PLD-derived PA production. Agonist-stimulated DRG generation can occur either through the PLD: phosphatidate phosphohydrolase pathway or via a direct PLC mechanism. Blocking PA formation with the alcohols should block DRG generation through the PLD pathway. This is indeed the case, and both ethanol and butanol have been shown to inhibit DRG production in a concentration-dependent manner (39). Obviously any DRG produced in the presence of high concentrations of the alcohols must be generated by an alternative mechanism, most probably via PLC.

This approach has been used to estimate the contribution of PLC to agonist-stimulated PC breakdown in cultured fibroblasts, endothelial cells, and smooth muscle cells (40). These studies have provided some evidence to support the contention that PLC contributes to agonist-stimulated DRG production in some cell types. Caution must be exercised when interpreting such observations, particularly since the effect of high alcohol concentrations on PLC enzymes is not clear. In human neutrophils PIP_2-specific PLC was not inhibited by butanol or ethanol at concentrations up to 30 mM and 200 mM, respectively (39). At the higher concentrations both alcohols augmented agonist-stimulated (IP_3) production. Furthermore, ethanol is reported to induce phosphatidylinositol and phosphatidylethanolamine hydrolysis in cells (41–43).

In theory, inhibitors of phosphatidate phosphohydrolase could also be used as probes to distinguish between PLC- and PLD-mediated DRG production. High concentrations of propanolol (200 mM) almost completely inhibit phosphatidate phosphohydrolase and block PLD-derived DRG production (44). The likelyhood of non-specific effects induced by very high concentrations of propanolol is great, and therefore its utility as a probe for monitoring PLC activation will be limited. These indirect assays can only provide circumstantial evidence for the activation of PLC in cells. The production of more conclusive evidence will ultimately depend upon the identification of potent and selective enzyme inhibitors.

Acknowledgements

The authors would like to thank Drs M. Wakelam and S. Cook for providing the data in *Figure 1*. We are also grateful to all our laboratory colleagues for providing the experimental details on which many of the Protocols described in this chapter are based.

References

1. Berridge, M. J. and Irvine, R. (1989). *Nature*, **341**, 197–205.
2. Billah, M. M. and Anthes, J. C. (1990). *Biochem. J.*, **269**, 281–91.
3. Bligh, E. G. and Dyer W. J. (1959). *Can. J. Biochem. Physiol.*, **37**, 911–17.
4. Dowdall, M. J., Barker, L. A., and Whittaker, V. P. (1972). *Biochem. J.*, **130**, 1081–94.
5. Cook, S. J. and Wakelam, M. J. O. (1989). *Biochem. J.*, **263**, 581–7.
6. Clarke, N. G. and Dawson, R. M. C. (1981). *Biochem. J.*, **195**, 301–6.
7. Yavin, E. (1976). *J. Biol. Chem.*, **251**, 1392–7.
8. Vance, D. E., Trip, E. M., and Paddon, H. B. (1980). *J. Biol. Chem.*, **255**, 1064–9.
9. Wang, F. L. and Haubrich, D. R. (1975). *Anal. Biochem.*, **63**, 195–201.
10. Choy, P. C., Whitehead, F. W., and Vance, D. E. (1978). *Can. J. Biochem.*, **56**, 831–5.
11. Truett, A. P., Snyderman, R., and Murray, J. J. (1989). *Biochem. J.*, **260**, 909–13.
12. Pai, J.-K., Siegel, M. I., Egan, R. W., and Billah, M. M. (1988). *Biochem. Biophys. Res. Commun.*, **150**, 355–64.
13. Augert, G., Bocckino, S. B., Blackmore, P. F., and Exton, J. H. (1989). *J. Biol. Chem.*, **264**, 21689–98.
14. Martin, T. W. (1988). *Biochim. Biophys. Acta*, **962**, 282–96.
15. Lapetina, E. G. and Seiss (1987). *Methods Enzymol.*, **141**, 176–92.
16. Mallows, R. S. E. and Bolton, T. B. (1987). *Biochem. J.*, **244**, 763–8.
17. Abdel-Latif, A. A., Owen, M. P., and Matheny, J. L. (1976). *Biochem. Pharmacol.*, **25**, 461–9.
18. Touchstone, J. C., Chen, J. C., and Beaver, K. M. (1979). *Lipids*, **15**, 61–2.
19. Bartlett, G. R. (1959). *J. Biol. Chem.*, **234**, 466–8.
20. Buss, J. E. and Stull, J. J. (1983). *Methods Enzymol.*, **97**, 7–15.
21. Bocckino, S. B., Wilson, P. B., and Exton, J. H. (1987). *FEBS Lett.*, **225**, 201–4.
22. Bocckino, S. B., Blackmore, P. F., Wilson, P. B., and Exton, J. H. (1987). *J. Biol. Chem.*, **262**, 15309–15.
23. Agwu, D. E., McPhail, L. C., Wykle, R. L., and McCall, C. E. (1989). *Biochem. Biophys. Res. Commun.*, **159**, 79–86.
24. Domino, S. E., Bocckino, S. B., and Garbers, D. L. (1989). *J. Biol. Chem.*, **264**, 9412–19.
25. Goppelt, M. and Resch, K. (1984). *Anal. Biochem.*, **140**, 152–6.
26. Nakamura, K. and Hanada, S. (1984). *Anal. Biochem.*, **142**, 406–10.
27. Eryomin, V. A. and Poznyakov, S. P. (1989). *Anal. Biochem.*, **180**, 186–91.

28. Rubin, R. (1988). *Biochem. Biophys. Acta*, **156**, 1090–6.
29. Korchak, H. M., Vosshall, L. B., Zagon, G., Ljubich, P., Rich, A. M., and Weissmann, G. (1988). *J. Biol. Chem.*, **263**, 11090–7.
30. Takuwa, Y. Takuwa, N., and Rasmussen, H. (1986). *J. Biol. Chem.*, **261**, 14670–5.
31. Hii, C. S. T., Kokke, Y. S., Pruimboom, W., and Murray, A. W. (1989). *FEBS Lett.*, **257**, 35–7.
32. Martinson, E. A., Goldstein, D., and Heller Brown, J. (1989). *J. Biol. Chem.*, **264**, 14748–54.
33. Daniel, L. W., Waite, M., and Wykle, R. L. (1986). *J. Biol. Chem.*, **261**, 9128–32.
34. Preiss, J., Loomis, C. R., Bishop, W. R., Stein, R., Niedel, J. E., and Bell, R. M. (1986). *J. Biol. Chem.*, **261**, 8597–600.
35. Bocckino, S. B., Blackmore, P. F., and Exton, J. H. (1985). *J. Biol. Chem.*, **260**, 14201–7.
36. Tyagi, S. R., Burnham, D. N., and Lambeth, J. D. (1989). *J. Biol. Chem.*, **264**, 12977–82.
37. Pai, J.-K., Siegel, M. I., Egan, R. W., and Billah, M. M. (1988). *J. Biol. Chem.*, **263**, 12472–7.
38. Randall, R. W., Bonser, R. W., Thompson, N. T., and Garland, L. G. (1990). *FEBS Lett.*, **264**, 87–90.
39. Bonser, R. W., Thompson, N. T., Randall, R. W., and Garland, L. G. (1989). *Biochem. J.*, **264**, 617–20.
40. Huang, C. and Cabot, M. C. (1990). *J. Biol. Chem.*, **265**, 14858–63.
41. Kiss, Z. and Anderson, W. B. (1989). *FEBS Lett.*, **257**, 45–8.
42. Hoek, J. B., Thomas, A. P., Rubin, R., and Rubin, E. (1987). *J. Biol. Chem.*, **262**, 682–91.
43. Simonsson, P., Ferencz, I., and Alling, C. (1989). *Drug Alcohol Depend.*, **24**, 169–74.
44. Billah, M. M., Eckel, S., Mullmann, T. J., Egan, R. W., and Siegel, M. I. (1987). *J. Biol. Chem.*, **264**, 17069–77.

7

The determination of phospholipase A$_2$ activity in stimulated cells

MICHAEL J. O. WAKELAM and SUSAN CURRIE

1. Introduction

Phospholipase A$_2$ catalyses the hydrolysis of the ester bond between a fatty acid and the hydroxyl group at the 2-position of the glycerol backbone of a phospholipid, releasing the fatty acid and generating a lysophospholipid. In the majority of cases, the activation of the enzyme in mammalian cells results in the release of arachidonic acid from the phospholipid and it is this reaction which provides the focus for this chapter. A number of phospholipids have been reported to be substrates for stimulated phospholipase A$_2$ activity including phosphatidylinositol (PtdIns) and phosphatidylcholine (PtdCho).

Phospholipase A$_2$ activity can be stimulated in cells by a range of treatments. It has been known for some time that artificial elevation of intracellular [Ca^{2+}] by the use of an ionophore such as A23187 (1) could stimulate the activity of the enzyme. There are also a number of reports in the literature demonstrating that phospholipase A$_2$ can be stimulated by activators of protein kinase C such as *sn*-1,2-diacylglycerol and 12-phorbol myristate 13-acetate (2); however, it has been suggested that at least part of this activation is due to a direct activation of the phospholipase and is not mediated via protein kinase C. Of particular interest are the recent reports suggesting that phospholipase A$_2$ activity can be stimulated in cells as a result of direct receptor activation and that this involves the input of a G-protein (3). This G-protein input has been proposed to be both of the 'classical' form involving an activated α subunit, but also to involve the direct effect of released βγ subunits (4).

Activation of phospholipase A$_2$ can have a number of consequences for the cell. Arachidonic acid is the precursor molecule for a number of important signalling pathways, the fatty acid can be utilized in the synthesis of both prostaglandins and leukotrienes via the 'cyclic' and 'linear' pathways of arachidonic acid metabolism, respectively (5, 6). Alternatively, arachidonic acid itself may play a key signalling role *per se*. The fatty acid has been shown to be a physiologically relevant activator of certain isozymes of protein kinase

C (7), to stimulate the release of Ca^{2+} from the inositol 1,4,5-trisphosphate-sensitive store (8) and to modulate the guanine nucleotide bound state of p21^{ras} by inhibiting the activity *in vitro* of the *ras* GTPase activating protein (9). A number of reports also exist in the literature of agonist actions for arachidonic acid, for example it has been shown to act as a co-mitogen for fibroblast cells (10).

2. Phospholipid analysis

The descriptions of methodology in this chapter focus upon the use of Swiss 3T3 cells. However, the protocols provided should be directly applicable to all adherent cell types and easily adaptable to studies upon cells in suspension. The phospholipids of a cell are in a state of metabolic turnover thus it is straightforward to radiolabel different parts of the molecule by inclusion of the appropriate radiolabelled precursor in the culture medium. A number of points need to be borne in mind when such radiolabelling protocols are used. In particular it is important that the labelling period is long enough for an equilibrium state to be reached.

Fibroblasts are incubated for increasing times with [^{3}H]arachidonic acid, at the end of the incubation period the medium is removed, the lipids extracted and separated by thin-layer chromatography (TLC) prior to determination of incorporation of the radiolabel by scintillation counting (see *Protocol 1*). The cells can also be labelled with phospholipid head group precursors for analysis of lysophospholipid generation and loss or breakdown of phospholipids.

Protocol 1. The radiolabelling of and analysis of phospholipids

A. *Cell incubation and lipid extraction*

1. Plate the cells into suitable plastic culture dishes, generally 24-well plates, and grow until approximately 75–80% confluent. Then, change the medium to one containing 1 μCi/ml [5,6,8,9,11,12,14,15-^{3}H]arachidonic acid (150–230 Ci/mmol) and incubate the cells for 12, 24, 36, 48, 60, and 72 h.

2. Remove the plates from the tissue culture incubator and perform the subsequent procedures with the plates being warmed in an aluminium block plate holder. Remove the incubation medium and wash the monolayers with 3 × 0.5 ml HBG (see *Table 1*).

3. Add 1 ml ice-cold methanol to each well, scrape the contents with a wax-filled pipette tip or a PTFE-covered spatula and transfer to a glass tube. Wash the well with a further 0.5 ml ice-cold methanol and add to the tube, together with 0.75 ml chloroform. Mix thoroughly, and allow to extract for 30 min at room temperature.

4. Add 0.75 ml each of chloroform and water, vortex mix hard and allow the phases to separate. Transfer 1 ml of the lower, organic phase to a 2-ml glass trident vial. Dry the samples under vacuum, the use of a centrifugal evaporator such as a Uniscience Univap is recommended.

5. Dissolve the lipids in chloroform:methanol (95:5) and use for TLC.

B. *Thin-layer chromatography*

1. Soak a 20 × 20 cm silica gel 60 plate in 1 mM EDTA and air-dry. Activate at 120 °C for 1 h before use.

2. Spot samples and standard lipids on to the plate in a minimum volume.

3. Develop (70–90 min) the plate in a paper lined tank pre-equilibrated with a solvent of chloroform:methanol:glacial acetic acid:water (75:45:3:1, by volume).

4. Air-dry the plate in a fume cupboard and locate the standards by exposure to iodine vapour in an enclosed chamber.

5. By reference to the migration of standards, mark the phospholipid spots of interest, remove the silica from the plate with a razor blade and transfer to a scintillation vial, add 3 ml of an appropriate scintillation fluid (for example, Optiphase 3), mix, and determine the radioactivity by scintillation counting. Standard R_f values on this system are: Sphingo-myelin 0.075; PtdCho 0.125; PtdIns 0.48; PtdEtn 0.56; PtdSer 0.825; PtdOH 0.91. Under the culture conditions used the cells will have been confluent from after about 20 h, thus plotting the incorporation of radioactivity against time will allow the determination of the equilibrium labelling time for each lipid. This protocol can also be used to determine which lipids are labelled with arachidonic acid in the cell line(s) utilized.

Table 1. Composition of HBG

Concentration	Grams per litre
NaCl (137 mM)	8
KCl (5.36 mM)	0.4
$MgSO_4 \cdot 7H_2O$ (1.66 mM)	0.2
$MgCl_2 \cdot 6H_2O$ (0.49 mM)	0.1
$CaCl_2 \cdot 6H_2O$ (1.26 mM)	0.276
NaH_2PO_4 (0.35 mM)	0.042
$NaHCO_3$ (4.166 mM)	0.35
Glucose (10 mM)	1.8
BSA (1%, w/v)	10

3. Determination of phospholipase A_2 activation

3.1 Arachidonate generation

Most studies reported in the literature have simply analysed the change in radioactivity associated with the incubation medium of cells and taken the results to reflect the activation of phospholipase A_2 and the generation of arachidonate. There are a number of reasons why this is not a valid method. First, the release of radioactivity from cells is not a true reflection of the rate of phospholipase A_2 activation since the reaction is occurring inside the cell and thus the product must cross the plasma membrane to be released from the cell. Second, since no analytical method is employed in these studies, it is not certain that what is being measured is indeed arachidonate rather than an alternative eicosanoid. It is also possible that cell stimulation could be resulting in the release of a pre-formed arachidonate metabolite rather than reflecting the stimulation of arachidonate release. Consequently, it is essential that both intra- and extracellular arachidonate are analysed and that the identity of the radiolabelled eicosanoid is determined.

Protocol 2. The stimulated release of arachidonate

1. Radiolabel the cells on 24-well plates to equilibrium with 1 µCi/ml [^{3}H]arachidonic acid as determined in *Protocol 1*; for Swiss 3T3 cells this requires 24 h.

2. Remove the plates from the tissue culture incubator and incubate upon a plate warmer at 37 °C. Aspirate the labelling medium and wash the cells three times with Hanks' buffered saline containing 1% (w/v) bovine serum albumin and 10 mM glucose (HBG) at 37 °C. HBG is defined in *Table 1*.

3. Stimulate the cells by the addition of the appropriate agonist in 200 µl HBG, incubate at 37 °C for the required time.

4. Aspirate the medium, add 1 ml ice-cold methanol/15 µl glacial acetic acid.

5. Scrape the cell debris off the plastic and transfer to a glass tube, wash the well with 0.5 ml methanol and transfer to the same tube, add 0.75 ml chloroform to each tube, mix samples and leave to extract at room temperature for 30 min.

6. Add 0.75 ml each of water and chloroform, mix well, allow phases to separate—this can be accelerated by low-speed centrifugation. Aspirate the upper phase and pipette 1.5 ml of the lower, organic phase into a glass vial and dry under vacuum. Dissolve the samples in 1 ml petroleum ether (60–80 °C) containing 4% (v/v) diethyl ether. Then, apply the samples to silicic acid columns and arachidonate isolated by adsorption chromatography (see *Protocol 3*).

3.1.1 Adsorption chromatography

Adsorption chromatography upon silicic acid permits the separation of different lipid types such as phospholipids, mono-, di-, and triglycerides, and fatty acids. Since the cell extracts generated in the experiments are labelled with [^{3}H]arachidonate, the fatty acid fraction can generally be taken to represent the arachidonic acid, however it is essential that preliminary experiments indeed confirm the chemical identity of the radiolabelled component.

The silicic acid column procedure described in this chapter has been adapted from the method of Hirsch and Ahrens (11). Silicic acid, 325 mesh, must be activated before use. Some companies supply the product pre-activated, alternatively the method of Hirsch and Ahrens (11) is used. 18 g silicic acid is loaded into a glass column and washed successively with 10 ml diethyl ether, 30 ml acetone/diethyl ether (1/1, v/v) and then 20 ml diethyl ether. The column is then slowly washed (2–3 h) with 150 ml petroleum ether (b.p. 60–70 °C), this slow wash is to ensure the complete removal of the dehydrating solvents, the silicic acid is then stored dry. The silicic acid columns are prepared and used as described in *Protocol 3*.

Protocol 3. Preparation of silicic acid columns

1. Prepare silicic acid columns by adding 0.5 g silicic acid to glass-wool-plugged Pasteur pipettes supported in a rack. Wash each column with 10 ml petroleum ether (60–80 °C) before use.

2. Apply the samples, dissolved in 1 ml petroleum ether: diethyl ether (96:4, v/v) (see *Protocol 2*), to the columns and allow the solvent to run to waste. Then wash the columns with 3 ml petroleum ether:diethyl ether (96:4, v/v).

3. Collect the arachidonic acid containing fraction into a scintillation vial by eluting with 3 ml diethyl ether. Add 10 ml of scintillation fluid and determine the radioactivity by liquid scintillation spectrometry.

When the silicic acid chromatography method is being developed in a laboratory, it is advisable that the efficiency of the separation and the composition of the fractions is checked by TLC. Samples, prepared as in *Protocol 2*, are spotted on to silica gel 60 plates and chromatographed using the upper phase of a mixture of ethyl acetate:glacial acetic acid:2,2,4-trimethylpentane:water (45:10:25:50; by volume). Standard lipids are run on the same plates and their locations determined either radiochemically or by iodine staining.

3.2 Lysophospholipid determination

In addition to stimulating the release of arachidonic acid, phospholipase A$_2$ activation generates a lysophospholipid. Determination of the generation of a lysolipid is essential if the hormone-stimulated response being studied is assumed to be phospholipase-A$_2$-mediated since arachidonic acid can theoretically be released from diacylglycerol by the action of diacylglycerol lipase. This pathway has to be considered because those hormones which activate phospholipase A$_2$ generally also stimulate the activation of phospholipase-C-catalysed phosphatidylinositol 4,5-bisphosphate hydrolysis, consequently the generation of arachidonic acid may not reflect the activation of phospholipase A$_2$ rather the sequential activation of phospholipase C and diacylglycerol lipase.

The detection of lysophospholipid generation is frequently difficult, since the cell will remove the molecule as fast as possible because lysolipids have detergent-like properties and will thus be detrimental to the cell membrane in which they are formed. A further consideration in the detection of lysophospholipid generation is the choice of radiolabelled precursor. It is not advisable to use fatty acid labelling for these experiments since a number of assumptions then have to be made concerning the acyl structure of the phospholipids hydrolysed by the action of the phospholipase A$_2$. Consequently, phospholipid head group labelling is the preferable method. Since the detection of lysolipids is somewhat difficult, it is advisable to determine from which phospholipid(s) arachidonic acid is/are being generated before the experiment is performed (see *Protocol 1* for details). In Swiss 3T3 cells, stimulated phospholipase A$_2$ activity appears to hydrolyse only phosphatidylcholine, thus labelling of the cells with [^{3}H]choline and the subsequent identification of [^{3}H]lysophosphatidylcholine is the method of choice, the procedure outlined can be directly adapted to the determination of other lysophospholipids simply by changing the radioactive precursor to, for example, [^{3}H]ethanolamine. The alternative procedure is to label the cells with [^{32}P]Pi; however, this does not permit equilibrium labelling.

Protocol 4. Measurement of [^{3}H]lysophospholipid production

1. Culture Swiss 3T3 cells in 24-well plates and label with 2μCi/ml [^{3}H]choline for 48 h. The cell density at the onset of labelling (approx. 70%) is such that at the end of the labelling period the cultures are both confluent and quiescent.

2. Remove the plates from the tissue culture incubator and perform the subsequent procedures with the plates being warmed in an aluminium block plate holder. Remove the incubation medium and wash the

monolayers with 3 × 0.5 ml HBG (see *Table 1*). Then stimulate the cells for the appropriate times by the addition of agonist in 200 μl HBG.

3. Remove the medium by aspiration and add 1 ml ice-cold methanol to each well, scrape the contents with a wax-filled pipette tip or a PTFE-covered spatula and transferred to a glass tube. Wash the well with a further 0.5 ml ice-cold methanol which is added to the tube, together with 0.75 ml chloroform. Mix the tubes thoroughly and allow to extract for 30 min at room temperature.

4. Add 0.75 ml each of chloroform and water, vortex mix hard and allow the phases to separate. Transfer 1 ml of the lower, organic phase to a 2-ml glass trident vial and dry the samples under vacuum.

5. Dissolve the lipids in chloroform:methanol (95:5) and use for TLC.

6. Soak a 20 × 20 cm silica gel 60 plate in 1 mM EDTA and air-dry; activate at 120 °C for 1 h before use.

7. Spot samples and standard lipids on to the plate in a minimum volume.

8. Develop the plate (70–90 min) in a tank pre-equilibrated with a solvent of chloroform:methanol:glacial acetic acid:water (50:30:8:3, by volume).

9. Air-dry the plate in a fume cupboard and locate the standards by exposure to iodine vapour in an enclosed chamber.

10. By reference to the migration of standards, locate the phospholipid spots of interest, remove the silica from the plate with a razor blade and transfer to a scintillation vial, add 3 ml of an appropriate scintillation fluid (such as Optiphase 3) and determine the radioactivity by scintillation counting. Standard R_f values on this system are: PtdCho 0.312; PtdEtn 0.79; PtdSer 0.79; PtdOH 0.825; lysoPtdCho 0.106; lysoPtdEtn 0.52; lysoPtdOH 0.60; lysoPtdIns 0.40; lysoPtdSer 0.306.

3.3 Arachidonate mass determination

Whilst the radiolabelling methods outlined above provide information as to the rate of phospholipase A_2 activation and from which phospholipid(s) the fatty acid is released, they cannot provide information as to the amount of arachidonate actually generated. This is particularly important when it is considered that the reported effects of arachidonate upon protein kinase C activity and the release of calcium from intracellular stores requires micromolar concentrations of the fatty acid. There are a number of possible methods whereby the mass of arachidonate can be measured including determination by gas liquid chromatography, but a convenient method involves conversion of the fatty acid to a metabolite whose concentration can be determined by radioimmunoassay.

Protocol 5. The determination of arachidonic acid mass

1. Culture Swiss 3T3 cells as above in 24-well plates. First wash with 3 × 0.5 ml HBG over a period of 45 min, stimulate with an agonist (for example, bombesin) for 20 sec, and terminate incubations by the addition of 1 ml ice-cold methanol.

2. Leave the plates on ice for 20–30 min and then scrape the contents and remove to a tube together with an 0.5 ml methanol wash. Add 0.75 ml chloroform, mix the tubes vigorously and leave to extract for 30 min at 4 °C.

3. Split the phases by the addition of 0.75 ml each of chloroform and water. Take 1 ml of the lower phase to fresh glass tube (such as a trident vial) and dry under vacuum.

4. Add 1 ml petroleum ether:diethyl ether (96:4) and isolate the arachidonate-containing fraction by silicic acid chromatography as described in *Protocol 3*.

5. Dry the arachidonate-containing fraction under vacuum and redissolve in 20 μl ethanol.

6. Carry out incubations of 29 μl 0.1 M phosphate buffer pH 7.4, 5 μl soybean 15-lipoxygenase and 6 μl arachidonate sample or standard at 37 °C for 30 min. Terminate the reactions by the addition of 5 μl 20 μM nordihydroguaiaretic acetic acid (NDGA) and 5 μl sodium borohydride (50 mg/ml). Use arachidonic acid standards in the range 0.08–3.125 ng/ml.

7. Determine the 15-HETE content of the samples generated using a radioimmunoassay kit available from Amersham International.

---

4. Arachidonate metabolite generation

The procedures outlined above are applicable to the examination of acute responses; however, in many cells arachidonic acid is a metabolically active species being the precursor lipid for prostaglandins and leukotrienes. Therefore, it is important that the generation of products of the cyclo-oxygenase and the lipoxygenase pathways are monitored. There are numerous methods in the literature, a number are given below which can be used depending upon the available equipment. A number of radioimmunoassay kits are also commercially available for measuring prostaglandins and leukotrienes.

Protocol 6. Cyclooxygenase products

1. Seed Swiss 3T3 cells into 6-well plates and label with 1 µCi/ml [^{3}H]arachidonic acid for 24 h until confluent and quiescent.

2. Transfer the plates to an aluminium heated block at 37 °C and wash three times in 1 ml HBG. Then stimulate with agonist in 1 ml HBG for the appropriate times.

3. Terminate the reaction by addition of 1 ml methanol. The methanol contains PGE_2, 6-keto $PGF_{1\alpha}$, PGD_2, and $PGF_{2\alpha}$ (each at 2 µg/ml) as carrier prostaglandins.

4. Scrape the contents of each well and transfer to small glass centrifuge tubes. Chill on ice for 20 min and then centrifuge at 100 g for 15 min at 4 °C.

5. Transfer 0.8 ml of the supernatant to a glass vial. At this stage samples can be stored at −80 °C under a nitrogen atmosphere before HPLC analysis.

High-performance liquid chromatography

1. Acidify the samples to pH 3.0 by the addition of 0.2 ml 0.1 M sodium citrate/0.1 M citric acid, pH 3.0. then add 0.2 ml orthophosphoric acid to generate the correct aqueous:organic ratio for HPLC. This gives a total volume of 1.2 ml.

2. Load samples by injections of 1 ml on to a 30 cm × 39 cm µ-Bondapak C-18 reverse phase HPLC column (Waters-Millipore, Northwich, Cheshire, UK). Prepare this column for use by washing with 32% acetonitrile:68% orthophosphoric acid (v/v).

3. Perform the separation according to the following program:

Event	Time (min)	Flow (ml/min)	% acetonitrile	% orthophosphoric acid
1	0–30	1.0	32	68
2	30–30.5	2.0	32	68
3	30.5–34	2.0	95	5
4	34–37	2.0	32	68
5	37–37.5	1.0	32	68

4. Collect fractions (0.5 ml) in scintillation vials. Add 4.5 ml of Optiphase scintillation fluid and determine the radioactivity associated with each fraction by liquid scintillation spectrometry. Arachidonic acid and prostaglandins are detected by comparison with elution profiles of standards.

Protocol 6. *Continued*

5. Prepare standards for chromatography by dissolving in acetonitrile (32%)/orthophosphoric acid (68%). These can be detected by UV absorbtion at 216 nm. The retention times are as follows: PGE$_2$ 21.35 min; PGD$_2$ 25.14 min; PDF$_{2\alpha}$ 18.08 min; 6-ketoPGF$_{1\alpha}$: 25.94 min; and arachidonic acid 44.90 min.

Protocol 7. Lipoxygenase products

1. Seed Swiss 3T3 cells on to 6-well plates and radiolabel with [^{3}H]arachidonic acid (1μCi/ml) for 24 h until confluent and quiescent. The cell washings and incubations are exactly as outlined in *Protocol 4*.

2. Terminate cell incubations by the addition of 1 ml methanol containing 5-HETE and LTB$_4$ (2 μg carrier/ml methanol). Scrape the contents of each well and transfer to small glass centrifuge tubes. Increase the methanol content of the stopped reactions to 70% and keep the samples at −20 °C for at least 20 min in order to precipitate the protein. Following centrifugation at 1000 *g* for 15 min at 4 °C, remove the supernatant and reduce the methanol content of the samples to 50% by the addition of water.

3. Perform high-performance liquid chromatography separation of the lipoxygenase products upon a 25 cm × 0.39 cm LiChrosorb C-18 column (Merck). Make 1 ml injections on to the column, or alternatively use an autosampler.

4. Elute products with methanol 68.7%/(methanol 20%/ammonium acetate (0.5%) 80%) 31.3% at a flow rate of 2 ml/min for 12 min. Then, increase the methanol fraction to 93.7% for 6.5 min. Complete the run by decreasing the methanol to 50% for 3 min. The length of the run including equilibration time is 21 min.

5. Compare samples with the elution profiles of standards. The retention times are as follows: LTB$_4$ 5.20 min, and 5-HETE 11.90 min; arachidonic acid is eluted after 18.85 min on this system.

---

5. The control of phospholipase A$_2$ activity

The methodology outlined above can, and has been, utilized to determine the activation of phospholipase A$_2$ in cells stimulated with an agonist, such as bombesin in Swiss 3T3 cells (12), or with non-physiological stimulation such as an ionophore or a phorbol ester. In order to determine the mechanism of activation of the enzyme, experiments are performed utilizing permeabilized cells. There are several methods of cell permeabilization, though in the authors' hands a number of these are not usable for the study of phospholipase A_2 activity, at least in cultured Swiss 3T3 cells, since non-specific phospholipid hydrolysis is stimulated by the manipulations. The most effective and reproducible method in our hands has proved to be permeabilization with streptolysin-O (13).

Protocol 8. Cell permeabilization

1. Label confluent, quiescent cells with [^{3}H]arachidonic acid in 24-well plates as in *Protocol 1*.

2. Remove the labelling medium and wash the cells with 3 × 0.5 ml HBG.

3. Add 0.5 ml permeabilization buffer containing 0.6 units/ml streptolysin-O to each well and allow the permeabilization to procede for 5 min at 37 °C. The composition of the permeabilization buffer is outlined in *Table 2*. The source of the streptolysin-O is important with that from Wellcome Diagnostics being found by the authors to be most active. The free calcium concentration of the permeabilization has been calculated to be 100 nM.

4. Wash the cells with permeabilization buffer (minus streptolysin-O) for 10 min.

5. Add agonists or other stimulants (for example, guanine nucleotides) in a volume of 0.25 ml for the required times.

6. Terminate incubations by the addition of 0.25 ml methanol.

7. Scrape wells and process the samples for arachidonate generation or phospholipid determination as in previous protocols.

Table 2. Permeabilization buffer

Concentration	Grams per litre
Hepes 20 mM pH 7.5	5.2
KCl 120 mM	8.94
MgCl$_2$ 6 mM	1.22
CaCl$_2$ 0.061 mM	0.013
KH$_2$PO$_4$ 2 mM	0.28
Na-ATP 2 mM	1.22
EGTA 0.1 mM	0.038
BSA 1% w/v	10.0

When establishing a permeabilization methodology, it is critical that the efficiency of the procedure is monitored. The holes generated by the streptolysin-O method are large enough to allow cytosolic proteins to leak out, consequently permeabilization can be monitored by determining for example lactate dehydrogenase release.

6. Conclusions

In this short chapter we have outlined a number of methods which have proved useful and successful in our hands. The list of methods is clearly incomplete and it may be, for instance, that the reader wishes to determine the activity of the enzyme *per se*, for this there are a number of assays which can be found in the literature and particularly in ref. 14. It is, however, important to note the limitations of such assays in that phospholipase A$_2$ appears to be present in all cells, and the methods outlined in this chapter are aimed at determining that activity which is stimulated upon agonist activation.

Acknowledgements

Work performed in the authors' laboratory is supported by grants from the Wellcome Trust and Fisons plc.

References

1. Kanterman, R. Y., Felder, C. C., Brenneman, D E., Ma, A. L., Fitzgerald, S., and Axelrod, J. (1990). *J. Neurochem.*, **54**, 1225–32.
2. Pfannkuche, H. J., Kaever, V., Gemsa, D., and Resch, K. (1989). *Biochem. J.*, **260**. 471–8.
3. Burch, R. M., Luini, A., and Axelrod, J. (1986). *Proc. Natl. Acad. Sci. USA*, **83**, 7201–5.
4. Axelrod, J., Burch, R. M., and Jelsema, C. L. (1988). *Trends Neurol. Sci.*, **11**, 117–23.
5. Irvine, R. F. (1982). *Biochem. J.*, **204**, 3–16.
6. Johnson, M., Carey, F., and McMillan, R. M. (1983). *Essays in Biochem*, **19**, 40–141.
7. Murakami, K. and Routtenberg, A. (1986). *J. Biol. Chem.*, **261**, 15424–9.
8. Chow, S. C. and Jondal, M. (1990). *J. Biol. Chem.*, **265**, 902–7.
9. Tsai, M. H., Yu, C. L., Wei, F. S., and Stacey, D. W. (1989). *Science*, **243**, 522–4.
10. Takuwa, N., Takuwa, Y., and Rasmussen, H. (1988). *J. Biol. Chem.*, **263**, 9738–45.
11. Hirsch, J. and Ahrens, E. H. (1958). *J. Lipid Chrom.*, **233**, 311–20.

Michael J. O. Wakelam and Susan Currie

12. Currie, S., Smith, G. L., Crichton, C. A., Jackson, C. J., Hallam. C., and Wakelam, M. J. O. (1992). *J. Biol. Chem.* (In press.)
13. Howell, T. W. and Gomperts, B. D. (1987). *Biochim. Biophys. Acta*, **927**, 177–83.
14. Dennis, E. A. (ed.) (1991). *Methods in enzymology*, vol. 197. Academic Press, San Diego, New York, London.

8

Electrophysiological approaches to G-protein function

I. McFADZEAN and D. A. BROWN

1. Introduction

Many neurotransmitter receptors are coupled to ion channels in excitable cell membranes. Whilst in a number of cases (for example, the nicotinic acetylcholine receptor), the receptor and ion channel are part of the same macromolecular complex, in the majority of cases the receptor and the ion channel are separate entities and the coupling between receptor and channel employs a further, second messenger molecule. In recent years it has become clear that G-proteins play a fundamental role in this scheme, either by interacting with a primary effector enzyme, such as phospholipase C, to liberate soluble second messengers which in turn alter ion-channel activity or, as more recent work has shown, by interacting more directly with the ion channel. The aim of this chapter is to illustrate how electrophysiological techniques have been used to elucidate such transduction pathways and in particular the role of G-proteins. It concerns itself with a description of methods unique to electrophysiological experiments. Where for example, purified G-protein subunits have been used in electrophysiological experiments, the methodology relating to the purification procedure is not covered in detail.

2. The patch-clamp technique

The single major advance which enabled electrophysiologists to become involved in the study of what are essentially intracellular signalling events was the development in the late 1970s of the patch-clamp technique (1). Whilst a full review of the technology and methodology involved in making patch-clamp recordings is outside the scope of this chapter (the reader is referred to two excellent textbooks on the subject; see refs 2 and 3) it is worth spending a little time on a few salient features relevant to the forthcoming discussion.

2.1 Gigaseal formation

Good patch-clamp recordings rely on the formation of a physically strong seal with high electrical resistance (gigaohms) between the glass micropipette and the membrane of the cell under study. This is important for two main reasons. First, the patch of membrane underlying the micropipette is isolated, both electrically and physically, from the rest of the cell membrane. Second, high-resistance seals reduce the amount of electrical current noise, thus increasing the signal-to-noise ratio of the recording and allowing resolution of small (1 pA) single-channel currents.

Gigaohm seals are best obtained by optimizing the recording conditions as follows:

(a) The cell surface should be clean and free of debris or extracellular matrix.

(b) Both the pipette filling solution and, to a lesser extent, the solution bathing the cells should be particle-free.

(c) The micropipettes should be 'fire-polished' (2).

The requirement for the cell surface to be clean and free from debris means that in most cases either cultured cells, which are washed several times before making a recording, or cells which have been dissociated acutely using digestive enzymes such as trypsin or collagenase, are used. Recently, patch-clamp recordings have been made from central neurons contained within thin (100 μm) brain slices (4) and it is likely that this technique will be used in future to study G-protein function in neurons.

Micropipettes are filled, using a syringe fitted with a polythene canula and a 0.2 μm syringe filter. Small-volume syringe filters are available from most good laboratory suppliers. Since the quantity of intracellular solution required to fill the tip of a micropipette is no more that 100 μl, the filling solution can be made up as stock and stored frozen in 1 ml aliquots.

2.2 Patch-clamp configurations

Once the pipette has sealed on to the cell, different manipulations allow the experimenter to produce different configurations of the patch-clamp technique (*Figure 1*). With the pipette still attached to the cell the experimenter can record single ion-channel currents from the patch of membrane underlying the pipette. This is the 'cell-attached' or 'on-cell' configuration. Pulling back on the micropipette will cause a patch of membrane to be excised from the cell. In some cases a vesicle of membrane is formed, but this can be ruptured to form a patch by exposing the electrode tip to air. The patch so formed is in the 'inside-out' configuration since the cytoplasmic (inside) face of the patch is now open to the bathing solution. Note that in each of the above configurations, the micropipette filling solution is in contact with the extracellular face of the patch and therefore must approximate a typical

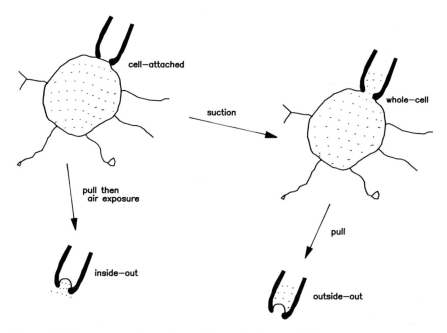

Figure 1. The various configurations of the patch-clamp technique.

extracellular solution in composition. If, when in the cell-attached configuration, the experimenter applies suction to the back of the micropipette then the area of membrane underlying the micropipette will be ruptured and the micropipette filling solution becomes contiguous with the cell cytoplasm. This is the 'whole-cell' configuration, and rather than recording single-channel currents the experimenter records ionic currents similar to those recorded using conventional voltage-clamp techniques.

The whole-cell method has, however, several advantages over conventional methods which rely on impaling the cell with sharp microelectrodes.

(a) It can be used on small cells (< 30 µm) which would be damaged irreversibly when impaled using even the sharpest microelectrodes.

(b) The micropipettes used to make whole-cell recordings are larger (tip diameter typically 1–2 µm) than those used to impale cells. This has two main consequences:

   i. The large internal diameter of the whole-cell micropipettes means that the micropipette filling solution dialyses freely into the cell cytoplasm, giving the experimenter some degree of control over the intracellular milieu. It has been shown that in bovine chromaffin cells (typically 20 µm diameter) small cations equilibrate within 10 sec of rupturing the patch (5). Larger molecules diffuse into the cell at a rate

169

dependent upon micropipette geometry and their molecular weight (6).

ii. The electrical resistances of the whole-cell micropipettes are typically an order of magnitude lower than those used to impale cells (5 MΩ compared to 50 MΩ). The lower series resistance increases the signal-to-noise ratio and improves voltage control during voltage-clamp experiments.

Ironically, the large diameter recording micropipettes also underlie the major disadvantage of the whole-cell clamp method: 'washout'. This arises when either an ionic current or an agonist response relies on the presence of a soluble intracellular constituent. Dialysis of the cell interior with the micropipette filling solution leads to the loss of this intracellular constituent and ultimately loss of the current or agonist response.

Washout can be reduced in three ways:

(a) Replace the intracellular constituent by including it in the micropipette filling solution. Obviously, this requires that the constituent be identified, which often occurs by a system of trial and error. The most common culprits are the nucleotides GTP (see below) and ATP. The ATP can be replaced either on its own (2–5 mM) or in combination with an ATP regenerating system; for example, 2 mM ATP plus 14 mM creatine phosphate and 50 U/ml creatine phosphokinase (7).

(b) Use smaller-diameter and therefore higher-resistance micropipettes, remembering that in doing so you sacrifice some degree of electrical control.

(c) Use the perforated patch (slow whole-cell) technique (8).

In the perforated-patch technique, rather than the patch of membrane underlying the micropipette being ruptured by suction, it is permeabilized by the inclusion of a pore-forming agent in the micropipette filling solution. The pores formed in the membrane are of a size sufficient to allow passage of ions and small molecules into or out of the cell, but not larger intracellular constituents. Currently, the most widely used membrane permeabilizing agent is the antifungal drug nystatin. This forms pores in the membrane with diameter around 0.8 nm, corresponding to a lower molecular weight cut off at around 200. The nystatin is dissolved in methanol (5 mg ml^{-1}) by sonication and added to the micropipette solution at a final concentration of 20–200 μg ml^{-1}.

Permeabilization of the membrane can be followed by monitoring the accompanying fall in series resistance, and usually occurs within 5 min of gigaseal formation. Some workers have found that the inclusion of nystatin in the micropipette reduces the success rate of gigaseal formation. To overcome this, the very tip of the micropipette can be filled with a nystatin-free solution, although it should be noted that this will increase the time taken for membrane permeabilization to occur, as the nystatin has to diffuse to the micropipette tip. Alternatively, it is possible to deliver the nystatin solution to the tip of the micropipette once the gigaseal has formed by means of a fine polyethylene tube inserted into the micropipette (8). This option is technically difficult and is usually resorted to only in extreme cases.

The fourth configuration of the patch-clamp technique, the 'outside-out' patch is achieved by pulling back on the micropipette whilst in the conventional whole-cell configuration. Assuming the cells are well-attached to the surface of the dish, this will rip off a small area of membrane, which will immediately reseal over the end of the micropipette, with the extracellular face in contact with the bathing solution. Note that in the whole-cell and outside-out configurations the micropipette filling solution is in contact with the cytoplasmic face of the membrane and must therefore approximate an intracellular solution in composition.

In the following discussion most experiments involve the use of patch-clamp techniques. Examples where conventional recording techniques have been used are also given.

Electrophysiological experiments to study G-protein function are in the main designed to answer three overlapping questions:

(a) Is a G-protein involved in the response to an agonist?

(b) Does G-protein activation lead to liberation of a soluble second messenger responsible for ion-channel modulation?

(c) Can the G-protein be identified?

3. Determining the involvement of a G-protein in an electrophysiological response

The first aim of many experiments is to determine simply whether a G-protein is involved in mediating the electrophysiological response to a given agonist. In most cases this involves the use of the whole-cell clamp method. It should be noted that none of the following experiments performed in isolation yields definitive evidence of G-protein involvement.

3.1 Is the response GTP-dependent?

Initial indications that a response might be mediated by a G-protein often arises from the finding that an agonist response is dependent upon intracellular GTP. The absence of GTP in the micropipette filling solution used to make whole-cell recordings can lead to washout of the response to agonists (see above). GTP-dependence can be tested in whole-cell clamp experiments by comparing responses obtained using micropipettes containing either 100 μM or no GTP (9).

Whilst the dependence of a response on intracellular GTP is good evidence for the involvement of a G-protein, the converse does not hold, i.e. GTP-independence does not necessarily mean absence of a G-protein link. There are two reasons for this. First, cells differ in their ability to generate GTP such that in cells with a high turnover, sufficient GTP is produced to maintain the response even when the recording micropipette does not contain the nucleotide. For example, it has been estimated that atrial myocytes maintain 25–50 μM internal GTP in the absence of added GTP in the micropipette solution (10). Second, the efficiency of intracellular dialysis varies, depending on both the cells under study and the geometry of the patch micropipette. Whilst it is possible to make theoretical calculations of the diffusion rate for molecules such as GTP from the tip of the micropipette (6) it is more difficult to predict the diffusion of such molecules within the intracellular milieu. These considerations are particularly important when using neurons with large processes which are probably not dialysed effectively even when using 'ideal' micropipettes.

3.2 The intracellular application of GTP/GDP analogues or aluminium fluoride

Having attempted to determine the GTP-dependence or otherwise of a response using the whole-cell patch technique, the next stage is to interfere directly with the G-protein cycle by introducing either guanine nucleotide analogues or AlF_3 into the cell.

3.2.1 GTP analogues

Non-hydrolysible analogues of GTP such as GTPγS and Gpp(NH)p are now used routinely to probe G-protein function (10–13). When using the whole-cell clamp method, the analogues are added to the micropipette filling solution at concentrations of 100 to 500 μM.* Stock filling solutions can be stored frozen in aliquots.

Assuming the involvement of a G-protein in a response, the use of GTPγS filled micropipettes has two related effects, depending upon the cells under study. In some cells, dialysis of the cell interior with the non-hydrolysable analogue leads to a time-dependent response even in the absence of agonist, i.e. receptor-independent, effects. Examples of this include inhibition of the m-current in cultured superior cervical ganglion cells (11) or activation of inwardly rectifying potassium channels in cardiac muscle cells (10). This phenomenon presumably reflects a resting turnover of the G-protein cycle in the absence of receptor activation. In other cell, no effect is seen until agonist is applied, but the agonist response then becomes irreversible. Such is the case for inhibition of the calcium current in cultured neuroblastoma × glioma

---

* Analogues of GTP or GDP often come as lithium salts; for example, the tetralithium salt of GTPγS. With 500 μM GTPγS in the recording electrode there is a potential intracellular concentration of 2 mM lithium. Since lithium may itself affect G-protein/receptor coupling (14), either lithium should be added to control solutions, or alternatively, a number of chemical houses will supply the sodium salts of the analogues.

hybrid (NG108–15) cells (12). An interesting aspect of the effect of GTPγS in these cells is that several applications of agonist, in this case noradrenaline, were required to produce a fully irreversible response. This might indicate either slow dialysis of the cell interior by the pipette filling solution or a high intracellular concentration of endogenous GTP. In cardiac myocytes the maximal, receptor-independent activation rate of potassium channels in the presence of intracellular non-hydrolysable analogues of GTP is dependent on the concentration ratio of the non-hydrolysable analogue to GTP rather than the absolute concentration of the analogue (10). The maximally effective ratio is dependent upon the analogue being used and is 5:1 for GTPγS and 20:1 for Gpp(NH)p. Since the endogenous GTP concentration is not always certain and may be as high as 50 µM, even in internally dialysed cells, negative results obtained using non-hydrolysable analogues are not always conclusive. More recently, differences in G-protein cycle turnover rate have led to differential rates of GTPγS action on different G-protein-mediated responses in the same cell (13).

A novel method of introducing GTP analogues into cells is to use 'caged' compounds (15). Here, 2 mM of the inactive, 1(2-nitrophenyl)ethyl P^3-ester derivatives of either GTPγS or Gpp(NH)p is included in the micropipette filling solution during whole-cell recordings. The active GTP analogues are released inside the cell by photolysis induced by a xenon arc flash, filtered using a 300–350 nm band pass filter. Each 0.5-msec-duration flash releases approximately 20 µM of the analogues with a half-time for release of less than 10 msec.

3.2.2 Aluminium fluoride

As an alternative to non-hydrolysable GTP analogues, AlF_3 can be applied intracellularly to activate G-proteins (16). In practice, this means the addition of 10 mM KF, CsF, or NaF to the micropipette filling solution. The F^- ions react with the aluminium in the glass of the micropipette to produce the active AlF_4^- species. Alternatively, 10 µM $AlCl_3$ can be used in combination with the fluoride salt.

3.2.3 GDPβS

GDPβS competes with GTP for the guanine nucleotide binding site of the G-protein to prevent its activation. Thus the intracellular application of GDPβS blocks agonist responses which rely on G-proteins. The whole-cell clamp method can again be used, with GDPβS being added to the pipette filling solution at a concentration of 100–500 µM (17). Effective dialysis of the cell interior by the micropipette filling solution is essential to see an effect of GDPβS, since the analogue is in competition with endogenous GTP. Thus, very high concentrations of GDPβS may be necessary (> 1 mM) and the lack of effect of GDPβS is in itself not good evidence for the lack of involvement of a G-protein.

Analogues of GTP and GDP can also be included in the filling solutions of high resistance electrodes used to make conventional intracellular recordings (18, 19). Diffusion into the cell from such electrodes is severely limited as indicated by the high concentrations of the analogues required, with 20 to 30 mM being typical. In large cells, for example, neurons of *Helix* or *Aplysia*, analogues can be applied by intracellular injection using relatively low resistance (30–50 MΩ when filled with 0.5 M KCl) micropipettes either by iontophoresis (13) or by pressure ejection (20). The analogues are dissolved in 120 mM KCl. Hepes (20 mM) may be added to maintain a pH of 7.4. Active ejection from sharp electrodes means that lower concentrations of analogue [for example, 0.1 mM GTPγS (20)] can be used.

4. Determining whether the transduction pathway involves a soluble second messenger

As mentioned above, G-protein activation can lead to changes in ion-channel activity following either a direct interaction between the G-protein and ion channel or the liberation of a soluble second messenger which interacts with the channel. The cell-attached, patch-clamp configuration can be used to examine the latter possibility by comparing the ability of neurotransmitter applied either inside or outside the patch electrode to alter single channel currents within the patch (21, 22). In the case of tight receptor ion-channel coupling, neurotransmitter applied outside the patch would have no effect. Recently, evidence for tight coupling between α-adrenoreceptors and calcium ion channels in sympathetic neurons was obtained in this way (21). Conclusive evidence for such a coupling mechanism involves the use of excised patches. In the study of receptor coupling to calcium channels this is not possible, since calcium channel activity quickly disappears in such patches. However, this approach has been used to good effect to elucidate the coupling between muscarinic receptors and inwardly rectifying potassium channels in cardiac muscle cells. Thus, muscarinic receptor agonists have no effect on the channel activity in cell-attached patches when applied outside the patch (22); more importantly, agonist, contained within the pipette-filling solution, is able to activate channels in inside-out membrane patches (23). This is dependent upon the presence of GTP in the solution bathing the cytoplasmic face.

5. G-protein identification

Having demonstrated the involvement of a G-protein in an electrophysio-logical response, the next stage is to identify the G-protein. Initial experiments aim to determine whether the G-protein is sensitive to pertussis toxin (Ptx).

5.1 Sensitivity to pertussis toxin

Ptx ADP ribosylates the α subunits of a number of G-proteins, and in doing so prevents receptor activation of the G-protein. To test for Ptx-sensitivity, *in vitro* preparations can be incubated with Ptx (200–500 ng ml^{-1}) for several hours. Alternatively, recordings can be made from preparations derived from animals pre-treated with the toxin. Thus electrophysiological responses to opioids are blocked both in NG108–15 cells pre-treated with 500 ng ml^{-1} Ptx for 4 h at 37 °C (12) and in neurons from *in vitro* slices of locus coeruleus prepared from rats injected with 1–1.2 μg of Ptx intracerebroventrically 1 to 5 days prior to making the recordings (24). An alternative method of intracellular application is to inject cells with pre-activated Ptx. This has been employed successfully in *Helix* neurons using a pressure ejection system (20). The toxin is pre-activated by incubating it at 20 °C for 30 min in a solution containing 25 mM dithiothreitol, 0.25 M NaCl and 0.05 M Na$_2$PO$_4$ (pH 7.0). Before being injected the toxin is diluted to 10 μg ml^{-1} in a solution of final composition 120 mM KCl, 5 mM NaCl, 0.002 mM Na$_2$PO$_4$, 0.5 mM dithiothreitol and 20 mM Hepes (pH 7.4). For control experiments, heat-inactivated (57 °C for 2 h) toxin can be used.

A number of G-proteins are Ptx-sensitive. More precise identification of the G-proteins involved in electrophysiological responses has relied on two techniques: namely, reconstitution experiments, and the use of G-protein specific antibodies.

5.2 Reconstitution experiments

Purified preparations of G-proteins or their subunits can be applied intracellularly, using either the whole-cell clamp technique (7, 25) or excised, inside-out patches (23, 26–28). The use of the whole-cell clamp method relies on the G-protein under study being Ptx-sensitive. This is because it is first necessary to inactivate endogenous G-proteins by pre-treating the cells with Ptx. Purified G-protein preparations, usually the α subunits, can then be tested for their ability to reconstitute the responses to agonists following their inclusion in the micropipette filling solution. Effective concentrations range from 15 to 100 nM, although it is important to note that the concentration within the micropipette is unlikely to reflect accurately equilibrium concentrations within the cell. This is supported by the observation that full reconstitution of agonist responses takes up to 30 min (25) presumably since it relies on diffusion of 40 kDa proteins into the cell.

Where the currents under study are stable in excised patches and the G-protein couples directly to the ion channel, G-proteins can be applied to the cytoplasmic face of inside-out patches. The most widely studied example is the activation of inwardly rectifying potassium channels by neurotransmitters in cardiac muscle cells (23, 26, 27) and, more recently, neurons (28). In these experiments, receptor-independent responses are obtained using G-protein α subunits pre-activated using a non-hydrolysable GTP-analogue. Unlike the situation with whole-cell clamp recordings, only pM concentrations of pre-activated α subunits are required to activate the channels.

The G-protein subunits are often solubilized using detergents. It is important to carry out controls to discount non-specific effects of the detergent on ion channel activity. Recently the zwitterionic detergent 3-[3-cholamidopropyl]-dimethyammoniol}-1-propanesulphonate (Chaps) has been shown to activate atrial potassium channels (26). This was previously used to solubilize the $\beta\gamma$ subunits of G-proteins and may have been responsible for the increased channel activity seen when the $\beta\gamma$ subunits were applied to excised patches (27). Interestingly, the $\beta\gamma$ subunits do appear to be capable of activating the inwardly rectifying potassium current but this effect is mediated by arachidonic acid metabolites (29, 30).

Purified preparations of G-proteins have been obtained either from brain tissue or erythrocytes. Since the different G-proteins are difficult to separate biochemically, experiments in which such preparations have been used are open to criticism when functional assignments are given to individual G-proteins. More recently, in an attempt to overcome this problem, α-subunits have been obtained from *Escherichia coli* transfected with the relevant cDNAs using a plasmid expression vector (23). However, these subunits are ten to fifteen times less effective than the purified α-subunits, for as yet unknown reasons.

5.3 Antibody studies

The main problem with reconstitution studies is that they do not positively identify the normally operating transduction pathway. To do this it is necessary to actively block the normal pathway. To this end, antibodies, raised against the α subunits of the different G-proteins, have been introduced into nerve cells with the aim of blocking receptor/G-protein coupling (20, 31).

The major difficulty in such experiments is getting the antibody into the cell in sufficient concentrations to produce its effect. One possibility is to include the antibody in the micropipette filling solution when carrying out whole-cell patch-clamp recordings. In the authors' experience this approach is not very satisfactory for three reasons. First, antibody binding is notoriously slow and might not reach an effective equilibrium during the time-course of a recording. Second, if the antibody is not purified, but in the form of an antiserum sample, then the inclusion of serum in the micropipette filling solution means that serum proteins can find their way on to the electrode tip, preventing gigaseal formation. Third, it is difficult to predict the true intracellular concentration of the antibody.

It is possible to overcome at least the first two problems mentioned above by loading the cells with antibody using a microinjection technique. The cells are first impaled with a sharp micropipette, injected with antiserum and then left to recover for several hours before making electrophysiological recordings. Controls cells are injected with pre-immune serum. The micropipettes used for making the injection are similar to those used for making conventional intracellular recordings, with DC resistances around 50 to 100 MΩ when filled with 3 M KCl. Before use, the micropipettes are broken back, using a piece of tissue paper, until the tip diameter is approximately 1 μm and a small droplet of antiserum can be seen to be ejected from the tip when it is submerged in liquid paraffin. The micropipette is clamped in a holder similar to those used for patch-clamp pipettes and ejection is achieved by applying positive pressure to the back of the micropipette, using a syringe. The injection pipette can be connected to a conventional bridge amplifier and used to record membrane potential. Penetration of the cell can therefore be monitored as a drop in membrane potential. This injection procedure has been used successfully in NG108–15 cells with around 70% of cells surviving the injection (31). It is doubtful if such a high success rate would be obtained in smaller neurons.

Once the cell has been injected, the micropipette is withdrawn, the position of the cell marked on the bottom of the dish, and a line drawing of the cell constructed to allow it to be relocated for recording at least an hour later. To check the efficiency of the injection procedure the antibody can be detected in a sample of cells using immunohistochemistry. In the study mentioned above, the antibodies were raised in rabbits against a synthetic decapeptide corresponding to the C-terminus of the α subunits of the respective G-proteins. The cells were plated on glass coverslips, injected, fixed at -20 °C with 95% ethanol, 5% acetic acid, before being incubated for 30 min at 37 °C with a 1:100 dilution of a fluoroscein-conjugated porcine antibody raised against rabbit immunoglobulins. The cells were then viewed under ultraviolet light.

In neurons of *Helix*, a roughly similar method has been used, with the following notable differences (20). First, purified, polyclonal antibody was injected at a concentration of 165 μg ml^{-1} in 120 mM KCl, 20 mM Hepes (pH 7.4). Second, for control experiments, heat-inactivated antibody (65 °C for 15 min) was used. Third, fast green (2 mg ml^{-1}) was added to the injected solution to verify injection into the cytoplasm. Finally, the injection electrode was left *in situ* throughout the experiment, typically lasting 20 min.

6. Conclusions

This chapter has aimed to illustrate how electrophysiological techniques, in particular the patch-clamp method, have been used to study the role of G-proteins in coupling neurotransmitter receptors to ion channels in excitable cell membranes. The intention was to appeal to both electrophysiologists and non-electrophysiologists alike. It is apparent from a number of the procedures outlined that some of the most fruitful experiments have involved collaborative ventures between electrophysiologists and workers in other disciplines. It is hoped that other collaborations might arise from an increased knowledge of what is possible, and that electrophysiology will continue to be a powerful tool in the study of G-protein function.

References

1. Hamill, O. P., Marty, A., Neher, E. Sakmann, B., and Sigworth, F. J. (1981). *Pflugers Arch.*, **391**, 85–100.
2. Sakmann, B. and Neher, E. (ed.) (1983). *Single channel recording.* Plenum Press, NY.
3. Standen, N. B., Gray, P. T. A. and Whitaker, M. J. (ed.) (1987). *Microelectrode techniques, the Plymouth Workshop manual.* The Company of Biologists Ltd. Cambridge, UK.
4. Edwards, F. A., Konnerth, A., Sakmann, B., and Takahashi, T. (1989). *Pflugers Arch.*, **414**, 600–12.
5. Marty, A. and Neher, E. (1981). In *Single channel recording* (ed. B. Sakmann and E. Neher). Plenum Press, New York.
6. Pusch, M. and Neher, E. (1988). *Pflugers Arch.*, **411**, 204–11.
7. Ewald, D. A. Pang, I-H, Sternweis, P. C., and Miller, R. J. (1989). *Neuron*, **2**, 1185–93.
8. Horn, R. and Marty, A. (1988). *J. Gen. Physiol.*, **92**, 145–59.
9. Pfaffinger, P. J., Martin, J. M., Hunter, D. D., Nathanson, N. M., and Hille, B. (1985). *Nature*, **317**, 536–8.
10. Breitwieser, G. E. and Szabo, G. (1988). *J. Gen. Physiol.*, **91**, 469–93.
11. Brown, D. A. (1988). *Trends in Neurosci.*, **11**, 294–9.
12. McFadzean, I. and Docherty, R. J. (1989). *Eur. J. Neurosci.*, **1**, 141–7.
13. Volterra, A. and Siegelbaum, S. A. (1988). *Proc. Natl. Acad. Sci. USA*, **85**, 7810–14.
14. Avissar, S., Schreiber, G., Danon, A., and Belmaker, R. H. (1988). *Nature*, **331**, 440–2.
15. Dolphin, A. C., Wooton, J. F., Scott, R. H., and Trentham, D. R. (1988). *Pflugers Arch.*, **411**, 628–36.
16. Loirand, G., Pacaud, P., Mironneau, C., and Mironneau, J. (1990). *J. Physiol.*, **428**, 517–29.
17. Holz, G. G., Rane, S. G. and Dunlap, K. (1986). *Nature*, **319**, 670–2.
18. Andrade, R., Malenka, R. C., and Nicoll, R. A. (1986). *Science*, **234**, 1261–5.
19. North, R. A., Williams, J. T., Surprenant, A., and Christie, M. J. (1987). *Proc. Natl. Acad. Sci. USA*, **84**, 5487–91.
20. Harris-Warrick, R. M., Hammond, C., Paupardin-Tritsch, D., Homburger, V., Rouot, B., Bockaert, J., and Gerschenfeld, H. M. (1988). *Neuron*, **1**, 27–32.
21. Lipscombe, D. Kongsamut, S., and Tsien, R. W. (1989). *Nature*, **340**, 639–642.

22. Soejima, M. and Noma, A. (1984). *Pflugers Arch.*, **400**, 424–31.
23. Yatani, A., Mattera, R., Codina, J., Graf, R., Okabe, K., Padrell, E., Iyengar, R., Brown, A. M., and Birnbaumer, L. (1988). *Nature*, **336**, 680–2.
24. Aghajanian, G. K. and Wong, Y. Y. (1986). *Brain Res.*, **371**, 390–4.
25. Hescheler, J., Rosenthal, W., Trautwein, W., and Schultz, G. (1987). *Nature*, **325**, 445–7.
26. Cerbai, E., Klockner, U., and Isenberg, G. (1988). *Science*, **240**, 1782–3.
27. Logothetis, D. E., Kurach, Y., Golpe, J., Clapham, D. E., and Neer, J. (1987). *Nature*, **325**, 321–6.
28. VanDongen, A. M. J., Codina, J., Olate, J., Mattera, R., Joho, R., Birnbaumer, L., and Brown, A. M. (1988). *Science*, **242**, 1433–7.
29. Kurachi, Y., Ito, H., Sugimoto, T., Shimizu, T., Miki, I., and Ui, M. (1989). *Nature*, **337**, 555–7.
30. Kim, D., Lewis, D. L., Graziadei, L., Neer, E. J., Bar-Sagi, D., and Clapham, D. E. (1989). *Nature*, **337**, 557–9.
31. McFadzean, I., Mullaney, I., Brown, D. A., and Milligan, G. (1989). *Neuron*, **3**, 177–82.

A1

Appendix
Suppliers of specialist items

Aldrich Chemical Company Ltd., The Old Brickyard, New Road, Gillingham, Dorset, SP8 4JL, UK.

Amersham International plc, Lincoln Place, Green End, Aylesbury, Buckinghamshire, HP20 2TP, UK.

Berthold UK Ltd., 35 High Street, Sawbridge, St. Albans, AL4 9DD, UK.

Bio-Rad Laboratories Ltd., Bio-Rad House, Maylands Avenue, Hemel Hempstead, Hertfordshire, HP2 7TD, UK.

Boehringer Mannheim UK Ltd., Bell Lane, Lewes, East Sussex, BN7 1LG, UK.

DuPont (UK) Ltd., NEN Products Division, Wedgewood Way, Stevenage, Hertfordshire, SG1 4QN, UK.

Lipidex Inc., 814 Embree Crescent, Westfield, NJ 07090, USA.

Lipid Products, Nutfield Nurseries, Crabhill Lane, South Nutfield, Redhill, Surrey, RH1 5PG, UK.

Luckham Ltd., Victoria Gardens, Burgess Hill, Sussex, RH15 9QN, UK.

Merck Ltd., Burnfield Avenue, Thornliebank, Glasgow G46 7TP, UK.

Pharmacia Ltd. (LKB), Davy Avenue, Knowlhill, Milton Keynes, Buckinghamshire, MK9 8PH, UK.

Savant, PO Box 3670, Fullerton, CA 92634, USA.

Sigma Ltd., Fancy Road, Poole, Dorset BH17 7NH, UK.

Simmons Moulding Ltd., Parkside, Coventry, Warwickshire, CV1 2NE, UK.

Whatman Labsales Ltd., St. Leonard's Road, 20/20 Maidstone, Kent, ME16 0LS, UK.

Zinnser Analytic, Howarth Road, Maidenhead, Kent, SL6 1AP, UK.

Index

Index

glucagon 8
glycerol labelling 131, 137, 144
G_o protein 18
  gel electrophoresis of 40–2
G_p protein 105
G-protein(s) 31–2, 75–6
  ADP-ribosylation of 28, 43–8, 69, 72, 175
  antibodies to 29, 176–7
  electrophysiological studies of 167–78
  functional assays of 31–54
  phospholipase A_2 and 153
  reconstitution of 6–8, 175–6
  see also individual proteins
G_s protein 75–6
  cholera toxin and 28, 43–8, 69, 72
  GTPase activity and 27, 38
  in vitro translated α subunit of 57–73
  mutant 69–72
  reconstitution of 7–8
GTPase 31
  assays of 26–7, 35–9, 54
  ras activating protein 154
GTP-azidoanilide 48–51, 54
guanosine diphosphate (GDP) 4, 31, 50
guanosine 5′-O-2-thiodiphosphate
  (GDPβS) 173
guanosine 5′-O-3-thiotriphosphate
  (GTPγS) 9, 51–3, 54
  in cyc⁻ membranes 67–8, 71
  in electrophysiology 172–3
guanosine triphosphate (GTP) 31
  agonist response and 4, 8, 9, 17, 25–6,
    171–2
  in cyc⁻ membranes 68–9, 71
  'GTP shifts' 9–15
  antagonist binding and 12
  guanylate cyclase assay with 101
  hydrolysis of, *see* GTPase
  photoreactive analogue of 48–51
guanylate cyclase 101–2
(5′-)guanylylimidodiphosphate (GppNHp) 9,
  10, 15
  in electrophysiology 172, 173
G_z protein 38–9

high-performance liquid chromatography
  (HPLC) 116, 140
homogenizers 6, 33

inositol lipids 107–8, 110–13
inositol monophosphatase 105
inositol phosphates 105–6
  analysis of 113–20
  extraction of 108–10
inositol polyphosphate 1-phosphatase 105
inositol tetrakisphosphate (IP$_4$) 105, 106
  analysis of 115, 116, 118–20

inositol trisphosphate (IP$_3$) 105, 106, 123
  analysis of 115–18, 119
iodoclonidine 16
isoproterenol 8, 9
  in cyc⁻ membranes 67–8, 69, 71

kinetics, agonist-binding 18–19, 23–6

leukotrienes 153, 162
lipid vesicles 6–7
lipoxygenase products 162
lithium 105, 113
lysophospholipid(s)
  phospholipase A_2 and 153, 158–9
  radiolabelled 132–3, 137, 144–5

magnesium 36, 50–1
membranes, *see* plasma membranes
modelling of receptor/G-protein interactions 21–6
muscarinic receptor(s) 17, 27, 31
  potassium channels and 174
  mutant G_sα subunit 69–72

N-ethylmaleimide (NEM) 29, 42
neutrophil chemotaxis receptor 27
nitric oxide 101

octyl-glucoside 7
opiate receptor(s) 12, 29

patch-clamp technique, principles of 167–71
perchloric acid 109
pertussis toxin 28, 43–8, 175
phosphate labelling 131–2, 144
phosphatidic acid (PA) 123
  assays of 130–7, 138–40, 149
phosphatidylalcohols 144–9
phosphatidylbutanol 144, 145–8
phosphatidylcholine (PC) 123, 149
  phospholipase A_2 and 153, 158–9
phosphatidylinositol 4,5-bisphosphate
  (PIP$_2$) 105, 123
phosphoinositidase C 31
phosphoinositides
  analysis of 110–13
  extraction of 107–8
phospholipase A_2 31, 153–64
phospholipase C (PLC) 105, 123, 149, 158
phospholipase D (PLD) 123, 149
  assay of 143–8
phosphorylcholine 123
  assays of 124–30